SELECTED WORKS OF ARCHITECTURAL DESIGN & RESEARCH INSTITUTE OF SOUTH CHINA UNIVERSITY OF TECHNOLOGY CO., LTD.

华南理工大学建筑设计研究院有限公司作品选（2009—2019）

丘建发　裴泽骏　编著

华南理工大学出版社
SOUTH CHINA UNIVERSITY OF TECHNOLOGY PRESS
·广州·

倪阳

华南理工大学建筑设计研究院有限公司 院长 / 总建筑师
全国工程勘察设计大师
博士生导师
哈佛大学 GSD 访问学者

序言

萧伯纳说过："人生不是一支短短的蜡烛，而是一支由我们暂时拿着的火炬，要把它燃烧得光明灿烂，然后交给下一代。"40年栉风沐雨，历任院长林克明、陈开庆、何镜堂以及众多老一辈华工设计院人肩负使命、树立航标，用实际行动探索并照亮了华工设计院产学研相结合的发展道路。这是坚守初心、励精图治铸品牌的40年，也是一脉相承、开拓创新的40年。而今，站在新的起点，我们正面临多元创新驱动的发展态势和现代化设计方式发展之变局。因此，我们仍需沿着岭南建筑之路，坚持产学研相结合，探索技术经济承包制改革，同时，布局新领域的拓展，实现多元发展目标，这是我们努力的方向和共同的梦想，我相信它就在我们的前方。

1953年，全国院系调整，华南工学院建筑系与基建处联合成立设计室。1958年，华南工学院建筑学系成立"建筑设计院"，这是华工设计院的前身。当时，林克明、陈伯齐、夏昌世、谭天宋等岭南现代建筑教育开拓者，提出"建筑教育必须结合设计实践"的理念，确立了设计实践与建筑教育相结合的办院基调。1979年，教育部正式发文批准建立华南工学院建筑设计研究院。伴随着改革开放的声声号角和国家经济建设高速发展的步伐，华工设计院扬帆起航。

1985年，岭南建筑大师莫伯治院士、佘畯南院士作为顾问教授开创设计院招收研究生的先河，促使设计实践与建筑教育结合得更加紧密，同时也让青年一代在入行之初就有了大师的悉心陪伴与专业指导，并由此逐步形成了华工设计院一脉相承的传习理念。

作为高校设计院，我们强调设计实践与建筑教育和学术研究相结合。以学科为依托，既是生产实践机构，为设计创新和实践提供平台；又承担培养研究生和技术研发的任务，为生产实践培育后备人才，反哺教学科研。这一理念在何镜堂院士担任院长后得到了进一步的发展，内涵也更加丰富。

为实现人才和物质资源的有效配置，并适应市场化的发展要求，何镜堂院长创新工作组织形式，在国内率先探索了"团队"运作模式，先后成立了何镜堂工作室及单专业工作室，并实行项目负责制。在该模式下的设计实践，既强调团队合作，在思想的碰撞中激活智慧，实现了方案优化和自我提升，又在研究领域和发展方向上各有侧重。

近些年来，我们培养了中国工程院院士1名、全国工程勘察设计大师4名、广东省工程勘察设计大师3名、当代中国百名建筑师4名、当代中国杰出工程师5名，中国建筑学会青年建筑师奖获奖者23名。基于此，一大批人才脱颖而出，成长为青年骨干、技术带头人乃至行业领军人。这些年来，我们创作了上海世博会中国馆、侵华日军南京大屠杀遇难同胞纪念馆、北京奥运会摔跤馆和羽毛球馆、汶川大地震映秀震中纪念馆、澳门大学横琴校区、钱学森图书馆、侵华日军731部队罪证陈列馆等一批具有国际影响力的作品，并在设计实践中树立了文化博览建筑、教育建筑、超高层建筑、体育建筑、会展建筑等几大设计品牌，大大提升了我院的影响力和竞争力。纵观我们走过的40年传承发展历史，我们以建筑空间记录、见证历史，并让我们自己成为历史的一部分。

何镜堂院长在设计实践中提出，做工程要做好三件事。第一件事是出一件作品，这是设计的直接结果；第二件事是写一篇文章，把它转化成理论与方法；第三件事是获一个奖，这是对作品的他证。这一理念对我在建筑创作上的影响非常大，我如今仍心存感激。对设计院这个整体来说，这一理念更是为我院树立了作品意识，为我院持续创作出高层次、高品位的建筑作品提供了指导。

截至2019年10月，我院累计获国家、部委和省市优秀设计奖900余项（国家级金奖5项、银奖6项、铜奖4项以及国际建筑大奖多项，国际著名网站ArchDaily中国年度大奖2项，等等），获奖数量同期居华南第一、全国前列。同时，我们还积极参与国家自然科学基金、"十三五"国家重点研发计划、省市科技计划的课题研发，从设计实践中发现问题，在研究过程中解决问题，并获得了国家科学技术奖、教育部科学技术奖、省市科学技术奖等科技奖励。不少研究成果或应用于设计实践，或提炼为行业标准，为推动行业科技进步贡献了力量。

本作品集出版于华工设计院成立40周年之际，书内挑选了本院2009-2019十年间具有代表性的项目，旨在总结经验，促进交流。

目录

CONTENTS

01 文化建筑 1
CULTURAL ARCHITECTURE

02 办公建筑 75
OFFICE ARCHITECTURE

03 教育建筑 109
EDUCATIONAL ARCHITECTURE

04 会展建筑 193
EXHIBITION ARCHITECTURE

05 体育建筑 211
SPORTS ARCHITECTURE

06 交通建筑 229
TRANSPORT ARCHITECTURE

07 其他建筑 249
OTHER ARCHITECTURE

08 城市设计 275
URBAN DESIGN

上海世博会中国馆	China Pavilion, Shanghai Expo 2010
世界反法西斯抗战胜利中国战区胜利纪念馆	Second World War China Victory Memorial Hall
侵华日军 731 部队罪证陈列馆	Invaded Japanese Army 731 Force Crimes Exhibition Hall
大厂民族宫	Dachang Cultural Center
长沙滨江文化园	Riverside Cultural Park in Changsha
国家南海博物馆	National South China Sea Museum
汶川大地震震中纪念馆	Wenchuan Earthquake Epicenter Memorial Hall
玉树博物馆	Yushu Museum
泰州民俗文化展示中心	Folk Culture Exhibition Center of Taizhou
惠能纪念堂	Huineng Memorial Hall
南京城墙博物馆改扩建工程	Expansion Project of Nanjing City Wall Museum
深圳市宝安中心区图书馆	Shenzhen Bao'an Centre Library
厦门海峡旅游服务中心	Xiamen Strait Tourist Service Centre
嘉善规划展览馆	Jiashan Urban Planning Exhibition Hall
天津博物馆	Tianjin Museum
三祖文化博物馆	Three Ancestors（Sanzu）Cultural Museum

01 文化建筑
CULTURAL ARCHITECTURE

烟台文化中心	Yantai Cultural Centre
梅州世界客商中心	The World Hakka Entrepreneurs Center
增城城市馆	Zengcheng City Gallery
广州古田村民中心	Guangzhou Gutian Villager Center
龙船山莞香文化公园二期——香馆工程	Longchuanshan Incense Cultural Park Phase ii — Incense Pavilion Project
广州粤剧艺术博物馆	The Museum of Cantonese Opera Art
北滘市民活动中心	Beijiao Civic Activity Center
广州国际残障人文化交流中心	Guangzhou International Disability Cultural Exchange Center
西关美食文化博物馆	Xiguan Food Culture Museum
南海会馆	Nanhai Guild Hall
迪拜清真寺	Scheme Design of Dubai Mosque
丽江市博物院新馆	Lijiang Museum
镇江市图书馆	Zhenjiang Library
崇左市文化艺术中心	Chongzuo Culture and Art Center
花山岩画游客服务中心	Huashan Rock Visitor Center

上海世博会中国馆

China Pavilion，Shanghai Expo 2010

项目位置：上海市
设计时间：2007 年
竣工时间：2010 年
建筑规模：72,480 m^2

Location：Shanghai
Design：2007
Completion：2010
GFA：72,480 m^2

BAOSTEEL
宝钢大舞台

中国馆的创作构思凝聚了全球华人的智慧与心血，体现了“东方之冠，鼎盛中华，天下粮仓，富庶百姓”的创作理念。

在整体布局上，中国馆居中升起，层叠出挑，庄严华美，形成凝聚中国元素、象征中国精神的主体造型“东方之冠”。地区馆水平展开，汇聚人流，以基座平台舒展形态衬托国家馆，展现出属于城市、面向世界的中国大舞台形象。

在场地设计上，整合南北城市绿地，形成坐南朝北、中轴领袖、大气恢宏的整体格局，体现了传统中国建筑与城市布局的经验与智慧。在技术设计上层层出挑的主体造型，显示了现代工程技术的力度美与结构美；对生态节能技术的综合运用，显示出我们对环境与能源等当今重大问题的关注与重视。

中国馆站在中国文化、东方哲学的立场上，对21世纪的城市与人居文明做出自己的诠释与展望。

The creative idea of China Pavilion embodies the wisdom and efforts of all Chinese people, symbolizing the concept of “the oriental crown, the splendid of China, the affluent barn, the affluent people”.

In the overall layout, the solemn national museum raises in the middle, symbolizing the spirit of China's main model “oriental crown” with cohesion Chinese elements. The regional museum spreads horizontally and gathers flows of people, which forms a base platform to serve as a foil for national museum displaying a China's great stage facing the word.

In the site design, the integration of the north and south urban green space forms the overall pattern of northward block, central axis leader and magnificent atmosphere, which reflects traditional Chinese experience and wisdom in architecture and urban layout. In the technical design, the layer-upon-layer form shows the beauty of modern engineering technology in strength and structure; meanwhile the comprehensive use on the ecological and energy-saving technologies demonstrates our concern and attention on environment and energy.

China Pavilion stands on the position of China's and prospect culture, oriental philosophy, and makes its own interpretation of human civilization and prospect in the 21st century.

4 / 5

世界反法西斯抗战胜利中国战区胜利纪念馆

Second World War China Victory Memorial Hall

项目位置：江苏省南京市
设计时间：2013 年
竣工时间：2015 年
建筑规模：54,980 m²

Location：Nanjing, Jiangsu
Design：2013
Completion：2015
GFA：54,980 m²

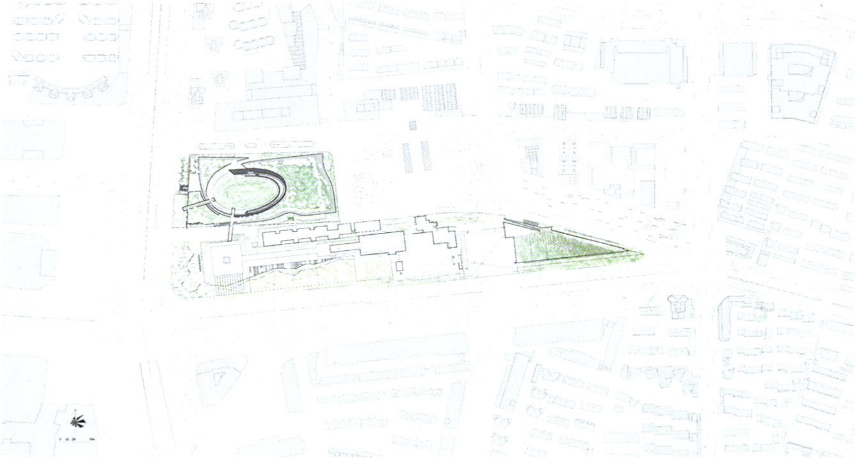

6 / 7

纪念馆的建设是为了纪念侵华日军1945年9月9日在南京投降事件，凸显抗战胜利的主题。构思立意为胜利、圆满的情感表达。胜利纪念馆是侵华日军南京大屠杀遇难同胞纪念馆扩建工程的一个补充和延续，兼具开放性与公共性，日常性与纪念性。这里是一个容纳历史记忆与当前生活、胜利喜悦与死亡悲痛的场所，人们可以在这里纪念、休息、放松、漫步、玩耍。我们希望通过一方公园、一个广场、一条道路的设计带给南京城市、南京市民喜欢的城市公共空间。

建筑利用绿色种植屋面、太阳能光伏发电、中水回收利用、透水混凝土选材、热压拔风效应等多种生态低碳措施，创造出一个节能环保的生态性纪念建筑。

The Memorial Hall was built to commemorate the surrender of the Japanese invaders in Nanjing on September 9, 1945, highlighting the theme of victory in the War of Resistance against Japanese Aggression. The idea is to express the feelings of victory and fulfillment. The Victory Memorial Hall is a complement and a continuation of the extension to the Memorial Hall of the Victims of the Nanjing Massacre by Japanese Invaders, characterized by openness, simplicity, and remembrance. This is a place to hold historical memories and current lives, triumphs and deaths; a place where people can commemorate, rest, relax, walk and play. We hope that the design of the park, the square and the road will bring better public urban spaces to Nanjing and its residents.

The building has adopted a variety of ecologically friendly, low-carbon measures. These include a green rooftop, a photovoltaic generator, gray water recycling, permeable concrete materials, and the hot air pressure and draft effect, to guarantee an energy-saving and environmentally-friendly memorial building.

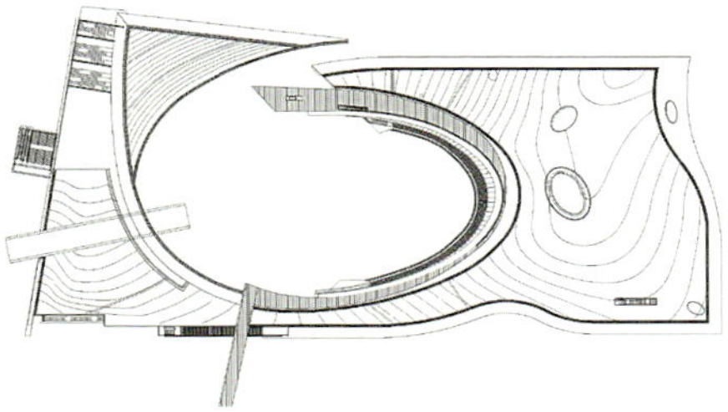

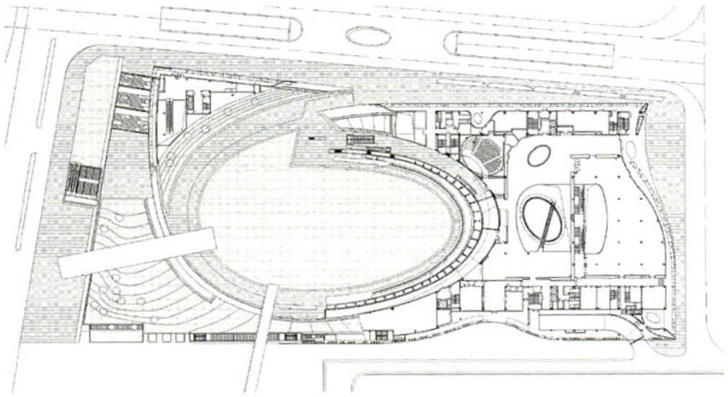

侵华日军731部队罪证陈列馆
Invaded Japanese Army 731 Force Crimes Exhibition Hall

项目位置：黑龙江省哈尔滨市
设计时间：2014 年
竣工时间：2015 年
建筑规模：9,997 m^2

Location：Harbin, Heilongjiang
Design：2014
Completion：2015
GFA：9,997 m^2

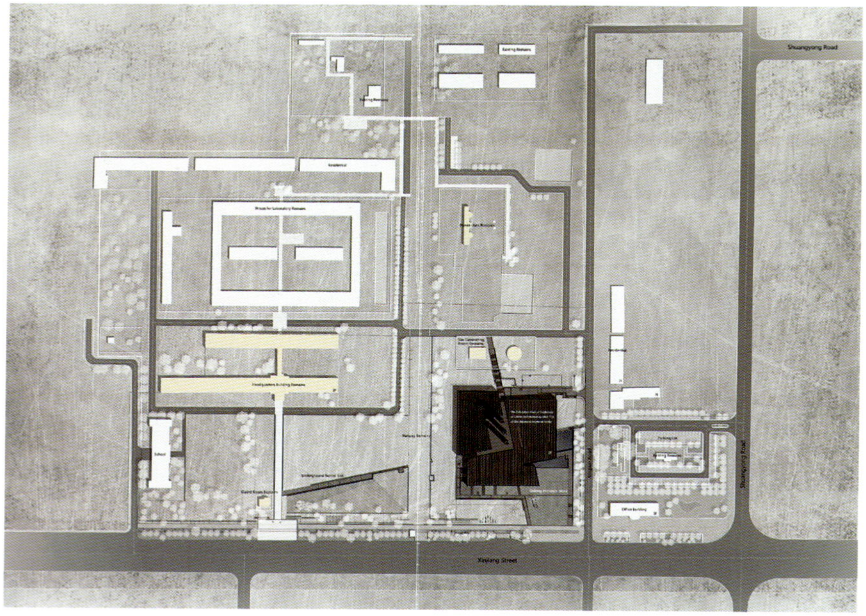

侵华日军 731 部队罪证陈列馆是为了保留细菌战罪证，并让人们铭记历史而建。项目建在现状遗址之上。1945 年，731 部队败逃之际炸毁了军事基地原有的大部分建筑，形成了现状遗址的基本格局。

项目的设计包含遗址场地设计和罪证陈列馆设计两大部分。场地设计方面，我们通过绿地公园过渡遗址与周边城市的空间关系。其他部分以保护原有遗迹为原则设计专用导游线路，同时恢复 731 部队时期原有路网及周边的围墙作为场地设计的记忆框架，有选择性地还原历史场景。方案外观灵感来源于飞机的“黑盒子”，象征着封存的历史证据。我们在建筑中设计了两条轴线，一轴指向陈列馆的主入口，另一轴指向动力锅炉房的烟囱，暗指受害者就是通过这样的流线经历生死过程。轴线交叉处设置三个烟囱与原址烟囱呼应。但两者含义不同，原来的烟囱是为了毁尸灭迹，而现在的烟囱则是作为采光井，使真相得以还原。总之，建筑的设计考虑都是从原有遗址出发，挖掘内涵，提取元素进行再现。

Invaded Janpanese Army 731th Force Crimes Exhibition Hall was designed to preserve the crime of bacteriological warfare and to allow people to remember history. The project is built on top of the existing ruins. In 1945, when the 731th Force fled, most of the original buildings of the military base were destroyed, forming the basic pattern of the current status.

The design of the project includes two parts: the design of the site and the design of the crime exhibition museum. In terms of site design, we built a spatial relationship with the surrounding cities through the transition site of the greenland park. In other parts, the special tour guide line was designed based on the principle of protecting the original remains. At the same time, the original road network and surrounding walls of the 731th Force were restored as the memory frame of the site design, and the historical scene was selectively restored. The appearance of the scheme was inspired by the “black box” of the aircraft, symbolizing the historical evidence of the storage. We designed two axes-one points to the main entrance of the 731 exhibition hall, and the other points to the chimney of the power boiler room, alluding to the victim's life and death process through such a streamline. Three chimneys are placed at the intersection of the axes to echo the original chimney. But the two have different meanings. The original chimney was used to destroy the corpse, and the current chimney is used as a light well to restore the truth. In short, the design considerations of the building are all based on the original site, excavating the connotation and extracting elements for reproduction.

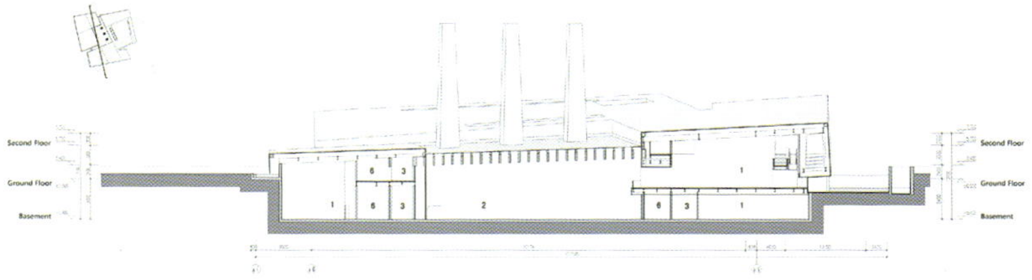

大厂民族宫
Dachang Cultural Center

项目位置：河北省廊坊市
设计时间：2014 年
竣工时间：2015 年
建筑规模：35,000 m^2

Location：Langfang, Hebei
Design：2014
Completion：2015
GFA：35,000 m^2

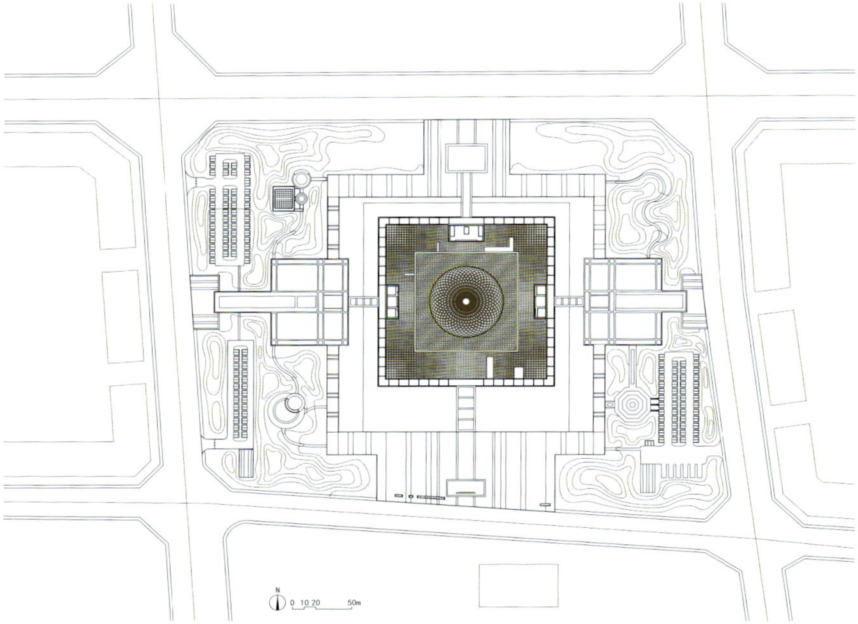

本项目建造在少数民族聚居区，是以弘扬民族文化、体现民族团结为主题，以博览和观演为主要功能的复合型文化综合体。

总体布局以“天圆地方”这一中国传统哲学和文化理念作为设计指导思想。建筑基座方正稳重，圆形主体高耸端庄，形成承天接地之势，恰如一颗汇聚和融合了中华各民族文化的璀璨明珠，镶嵌在大厂这片土地之上。建筑细节上汲取大厂当地回族建筑文化，与现代的建筑手法相结合，彰显精致、秀美的艺术气质，实现现代与传统的融合。精美、隽秀的花瓣形门洞形成柱廊空间；穹顶表面变换的花瓣形透空洞口，当阳光洒入，光影渲染出空灵和神秘的氛围。建筑与自然、光影融为一体。

Dachang Cultural Center is a cultural complex integrating the functions of theater, exhibition, convention and community center. We hope it will play a role in promoting national culture, reflecting ethnic unity and enriching the cultural life of local people of various ethnic groups.

The overall layout has been conceived to express the Chinese cultural connotation of a “round heaven and square earth”. Using traditional mosque designs combined with novel materials and techniques, the building has made good use of the spatial structure of the mosque.

长沙滨江文化园

Riverside Cultural Park in Changsha

项目位置：湖南省长沙市
设计时间：2005—2011 年
竣工时间：2014 年
建筑规模：总建筑面积：149,943 m^2；其中：博物馆 34,173 m^2，音乐厅 28,161 m^2，图书馆 31,322 m^2，规划展览馆 9,255 m^2

Location：Changsha, Hunan
Design：2005–2011
Completion：2014
GFA：Total construction area: 149,943 m^2; including Museum(34,173 m^2), Concert Hall (28,161 m^2), Library(31,322 m^2), City Planning Exhibition Hall(9,255 m^2)

长沙滨江文化园包括博物馆和城市规划展览馆、图书馆、音乐厅。项目设计有以下理念：

1. 尊重城市文脉

项目位于湘江与浏阳河交汇的新河三角洲，“顽石”和“沙洲”作为创作的构思原型，将大地景观元素融入城市形态之中，重塑了新河三角洲的大地肌理与地表形态。

2. 隐喻城市精神

建筑向洲头奔去的动势以及从大地中崛起的形象，象征湖湘文化勇往直前、倔强不屈的性格特点，通过整体建筑形态的隐喻与城市精神形成呼应。

3. 营造城市场所

设计以“开放共享，交流舞台”为出发点，为市民文化生活打造一个城市客厅。

Riverside Cultural Park in Changsha including: Museum and City Planning Exhibition Hall, Library, Concert Hall. The design includes the following concepts:

I Respect urban context

Project located on the Xinhe delta where the Xiangjiang river and Liuyang river converge, the shape derived from insensate stone and sandbank, brings architecture as an element of earth landscape, makes a harmonious coexistence with the environment.

II Illustrate city spirit

The architecture heads to delta and the tower rises from earth, symbolize the brave and unyielding character in Hunan culture.

III Create a city place

As a new landmark of Changsha culture, the design permeates into city as an open public space, provides an interchange culture activity place for citizens.

国家南海博物馆
National South China Sea Museum

项目位置：海南省琼海市
设计时间：2015 年
竣工时间：2017 年
建筑规模：70,870 m^2

Location：Qionghai, Hainan
Design：2015
Completion：2017
GFA：70,870 m^2

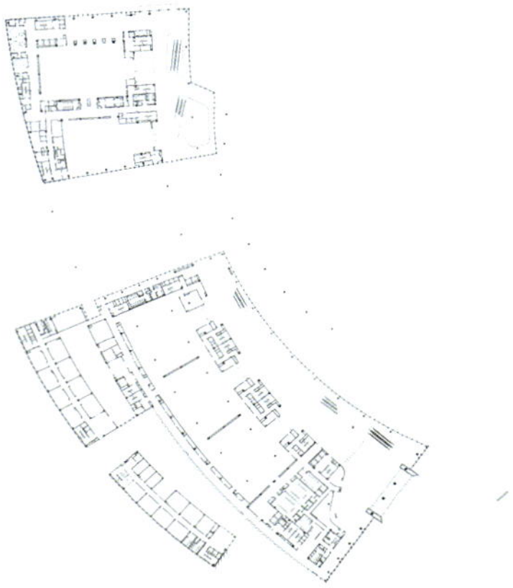

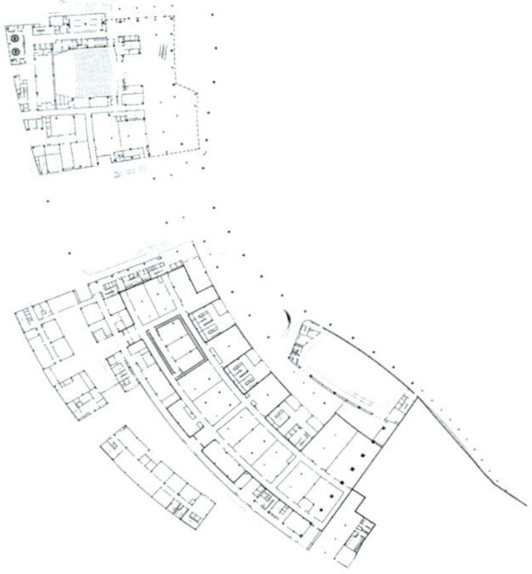

国家南海博物馆位于海南省琼海市潭门镇，是以海洋和南海文化为主题的大型国家级公益性文化设施，是体验南海文化、海南文化与海洋文化的现代博物馆与“一带一路”会展场地。

建筑内部形成张弛有序的、丰富的空间关系，满足大型文化建筑室内空间的层次性、明确性和高品质要求，空间设置与陈展内容相得益彰。从船屋发展而来的建筑形态，在通风、遮阳、隔热等绿色生态节能方面具有较为突出的优势。从建筑体型、室外景观、通风廊道、立面遮阳等多层面进行优化，满足海南的气候适应性要求，达到节能的目的。

Located in Tanmen Town, part of Qionghai in Hainan Province, the National South China Sea Museum is a large scale national public welfare cultural facility themed around the ocean and the South China Sea. It is a modern museum for experiencing the culture of Hainan and the South China Sea, and also showcasing marine life and traditions, as well as being a showcase for the “the Belt and Road” initiative.

The building interior features a relaxed, orderly, and enriched space. It can meet the requirements of hierarchy, clarity and high quality on the interior space of large cultural buildings, and its space setting and the content of the exhibition complete each other. The building body developed from the maritime-style housing has a number of advantages in terms of energy saving, including ventilation, shading, and heat insulation. The multi-level optimization of the building shape, outdoor landscape, ventilation corridor and shading allows the building to be well adapted to the climate of Hainan and to ensure an excellent level of energy saving.

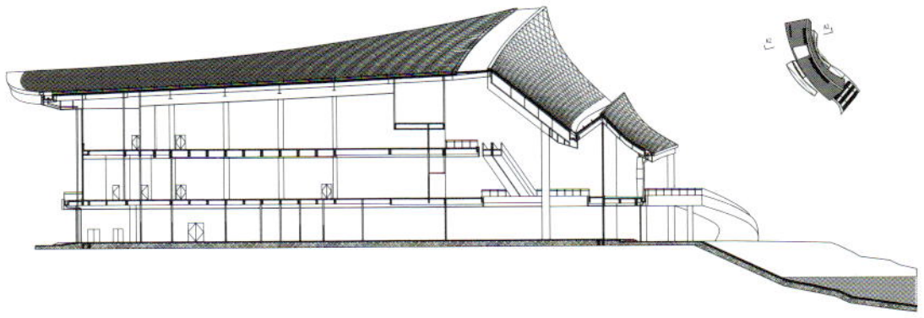

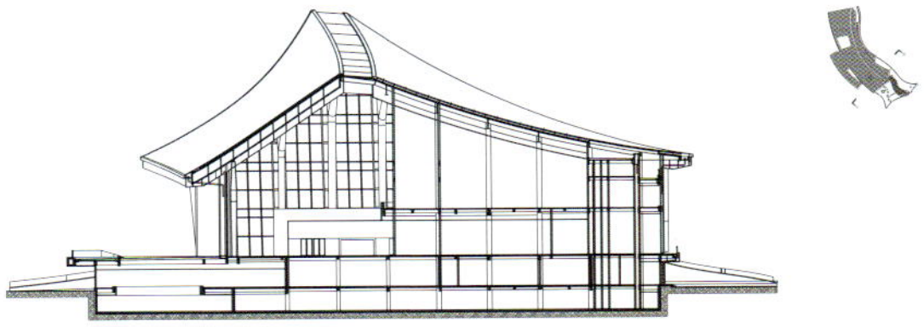

汶川大地震震中纪念馆
Wenchuan Earthquake Epicenter Memorial Hall

项目位置：四川省汶川县
设计时间：2009 年
竣工时间：2011 年
建筑规模：5,148 m^2

Location：Wenchuan, Sichuan
Design：2009
Completion：2011
GFA：5,148 m^2

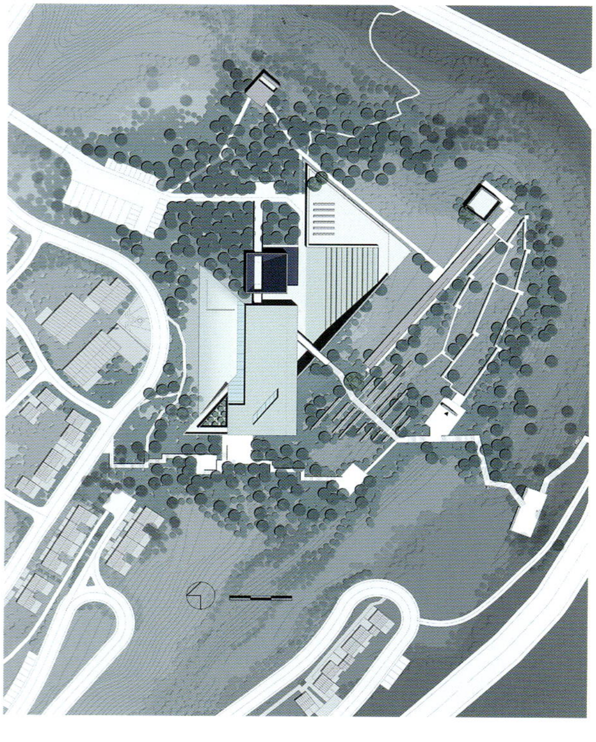

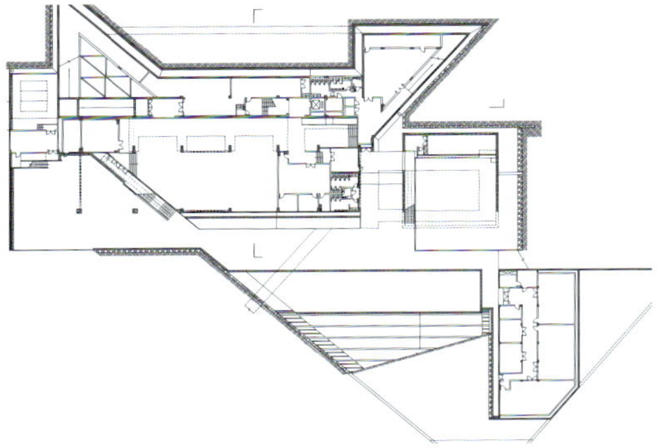

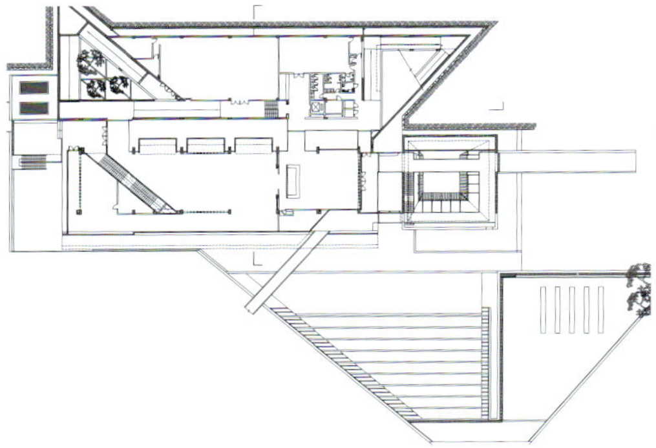

纪念馆位于2008年5月12日汶川大地震的震中映秀镇。它在高于城镇近60 m的山岗上，是突出的视觉焦点。它没有采用突兀的庞大体量，而是将最主要的建筑体量嵌入当地独特的山地景观，以低俯的姿态与地形紧密结合，平静而有力，质朴而厚重。

以流水、地殇、崛起、希望为主题的四个庭院，分别以水、石、树、光为构成要素营造场所精神。它们成为地景建筑与自然对话的窗口，人们在参观过程中与山体、天空、大地形成多次不同方式对话。

在参观流线上，设计将空间序列的组织、场所精神的营造与展览的主题内容密切关联。人们在空间的下沉、上升、开合、收放的节奏中，情绪由凝重、悲伤逐渐走向舒缓、希望。

Wenchuan Earthquake Epicenter Memorial Hall is located in Yingxiu, the epicenter of Wenchuan earthquake on May 12, 2008, nearly 60 meters above town on the hill. It is a prominent visual focus. Its size is not abrupt, and the main building volume is embedded into the local unique mountain landscape, closely integrated with low prone posture and terrain, which is calm and powerful, simple and heavy.

There are four courtyards with the theme of running water, ground war, rising and hope, respectively creating the site spirit with water, stone, tree and light as the elements. They are the windows of the landscape architecture and the natural dialogue, which can achieve many different ways of dialogue with the mountain, the sky and the earth during the visit.

As far as the visiting flow is concerned, the organization of space sequence and the construction of place spirit are closely related to the theme content of the exhibition, which make people's feelings move from dignified, sad to soothing and hopeful in the rhythm of the subsidence, opening and closing of space.

玉树博物馆
Yushu Museum

项目位置：青海省玉树自治州
设计时间：2011—2012 年
竣工时间：2014 年
建筑规模：11,500 m^2

Location：Yushu, Qinghai
Design：2011–2012
Completion：2014
GFA：11,500 m^2

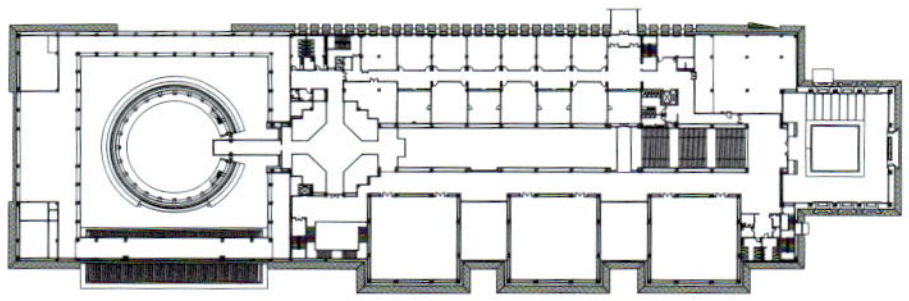

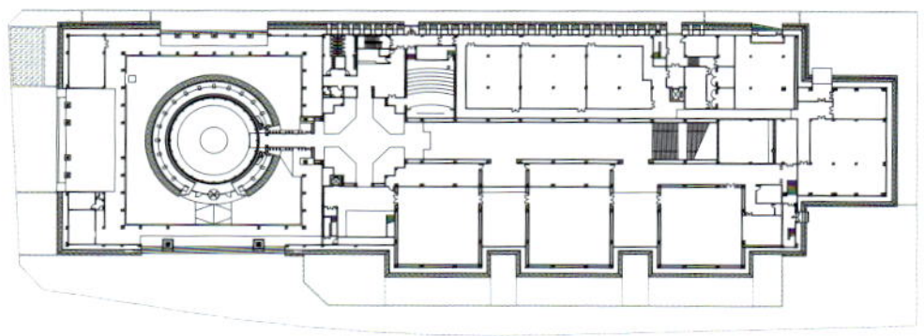

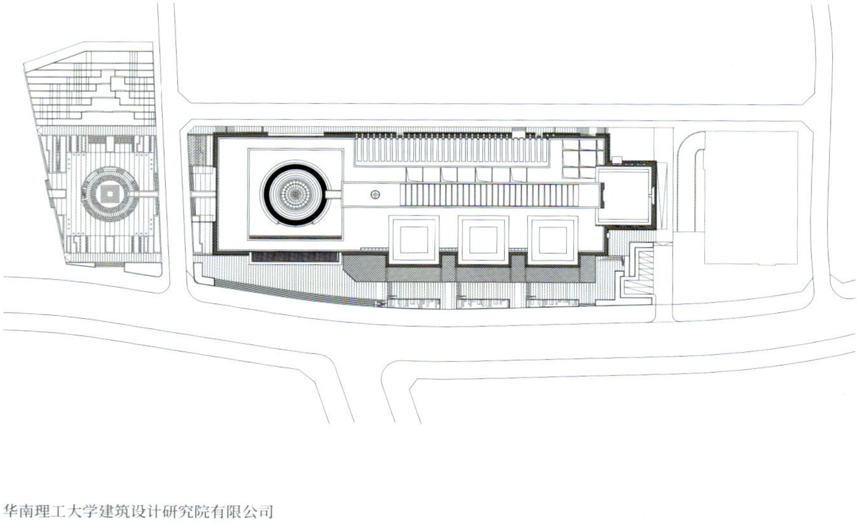

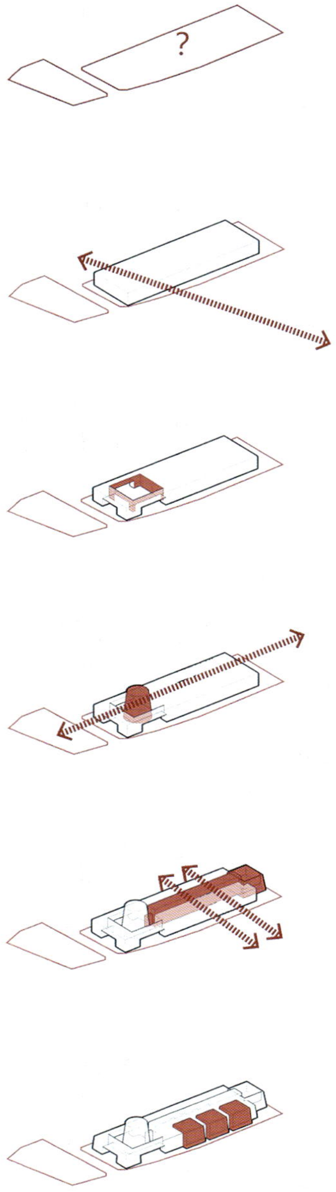

基地位于传统古街区，东接城市历史水脉“稻河头”，西邻名人旧居“多儿巷一号”，北靠小尺度的传统街巷肌理，南面为大尺度的现代城市体量。我们不仅关注建筑本身，而且把设计建造建筑范例与更新激活城市街区充分结合起来，营造融入传统记忆而又焕然一新的当代城市生活空间。

设计从理清和重建场地结构开始：延续和强化原有的历史水脉，建立指向名人旧居的东西向轴线，以新建展示馆为主体，从东到西带动起一系列具有叙事结构的城市空间序列，从北到南顺接了从历史街区过渡到现代城市的大小肌理与体量。毗邻而立的传统建筑成为建筑形态的源泉，设计注重以现代建造技术转译泰州传统建筑的精髓。

Located in a traditional ancient area, the center connects the city's historical water vein–Daohetou (Rice River Head)–in the east, neighbors the former residence of a local celebrity –Duo'er Lane No.1–in the west, makes use of small scale traditional street styles in the north, and faces the large scale modern city in the south. We not only paid attention to the architecture itself, but also fully combined the design and construction with the renewal and activation of urban blocks, so as to create a contemporary urban living space which integrates traditional memory with a new look.

We started the design by sorting out and modifying the structure of the site: continuing and strengthening the original historical water , and establishing an east-west axis pointed at the former residence of the local celebrity. The main exhibition hall forms the main body of the center , so as to create a series of urban spatial sequences with narrative structures running east to west , and to transit from the historical area to variously sized urban structures running north to south. The traditional buildings adjacent to each other became the source of the architectural forms used, and the design focused on modern construction technology to translate the essence of traditional Taizhou architecture.

泰州民俗文化展示中心

Folk Culture Exhibition Center of Taizhou

项目位置：江苏省泰州市
设计时间：2009 年
竣工时间：2014 年
建筑规模：11,397 m^2

Location：Taizhou, Jiangsu
Design：2009
Completion：2014
GFA：11,397 m^2

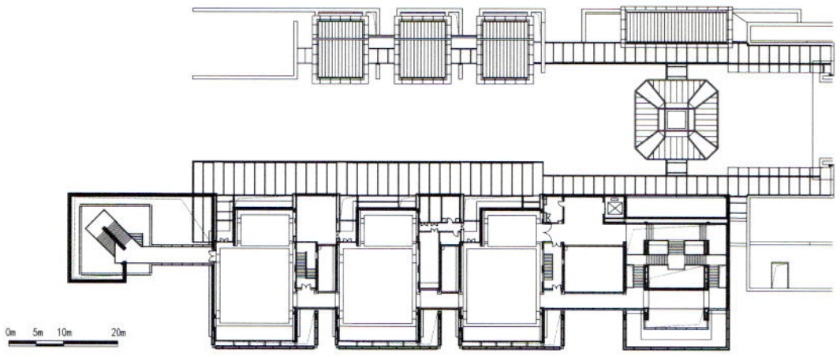

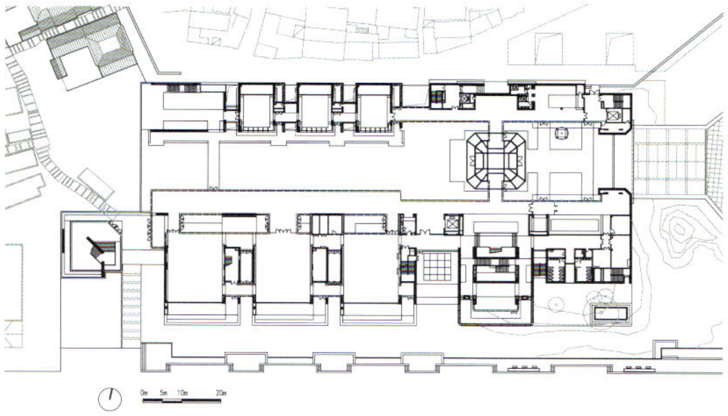

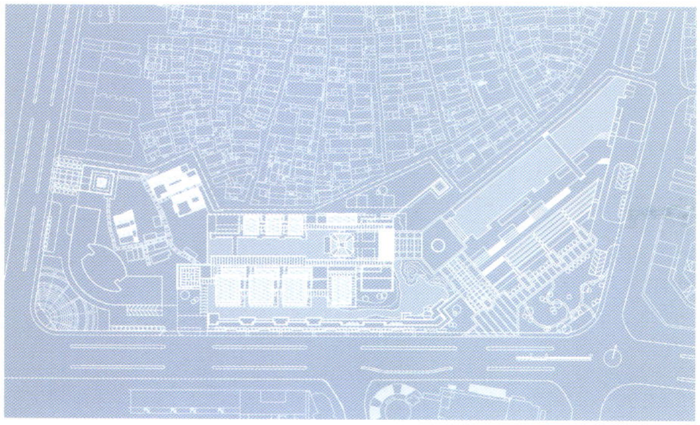

建筑被置于城市的关键节点，高低错落的台式体量融入城市肌理，“转经筒”以高昂的姿态体现藏族文化特征，重塑城市公共生活中心。

玉树州是青藏高原上的藏族文化名城。玉树博物馆是震后重建的重点项目之一。

基地位于三条河流和主要道路的交会处，是城市的核心地带。建筑以高耸的圆台式的形体作为城市两条轴线的焦点，成为城市空间的重要地标。我们围绕圆台设置了开放的院落，人们可以自由地穿越、逗留，具有强烈的公共性。

The building is located at a key hub of the city, with high and low platforms integrated into the urban texture. The “prayer wheel” reflects Tibetan cultural characteristics and reshapes the center as a space for urban public life.

Yushu is a famous Tibetan cultural city on the Qinghai-Tibet Plateau. Yushu Museum is one of the key projects for reconstruction after the earthquake.

The site is located at the intersection of three rivers and major roads as the core of the city. The building takes the towering round table form as the focus of the two axes of the city, which becomes an important landmark of the urban space. We set up an open courtyard around the round table, where people can freely cross and stay, with a strong publicity.

惠能纪念堂
Huineng Memorial Hall

项目位置：广东省云浮市
设计时间：2012 年
竣工时间：2013 年
建筑规模：10,842 m^2

Location：Yunfu, Guangdong
Design：2012
Completion：2013
GFA：10,842 m^2

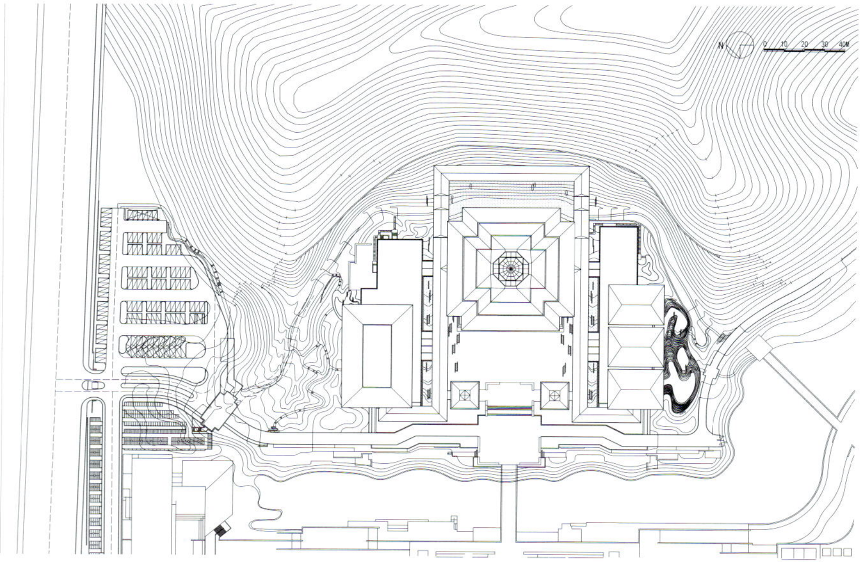

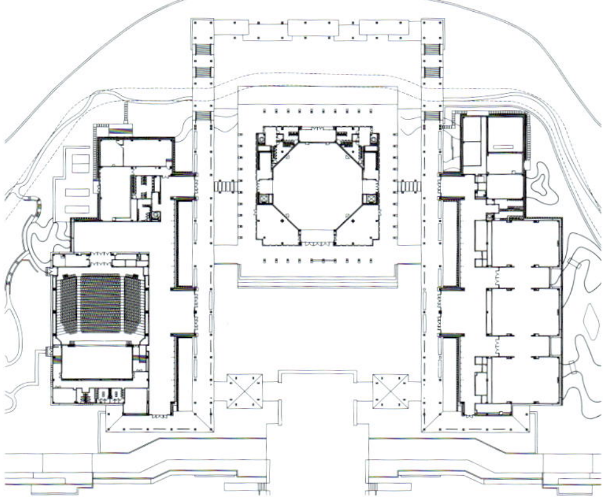

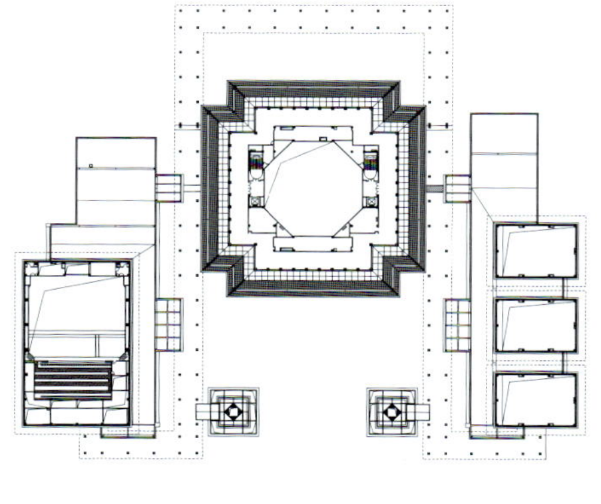

惠能纪念堂通过融于山水的场所、层层递进的院落、简洁纯净的空间，将禅宗精神与建筑艺术紧密结合，营造空明出世的禅宗圣境。基地中浅丘起伏，山塘延绵，层层叠叠的地形形成景深印象。惠能纪念堂融于山水之间，二者相得益彰——以反宇飞檐的重檐唐风形象与延绵的浅丘对话。建筑汲取传统佛教建筑的空间布局，通过空间收放组合，将内外庭院和主体大厅有机地串联在一起，层层递进，塑造典雅平和的禅意空间。

Huineng Memorial Hall surrounded by mountains and rivers, along with layers of progressive courtyards and pure, concise spaces, the spirit of Zen Buddhism is combined with architectural art, generating an atmosphere for a sacred land born out of otherworldliness. At its base, the rising and falling hillocks, majestic ponds, and layered terrain all give an intense feeling of perspective. The Huineng Memorial Hall is integrated into the mountains and rivers, with these two complementing each other. A dialogue is thus established with long stretches of shallow hillocks and Tang Dynasty style overhanging and dual eaves. Drawing on the layout designs of traditional Buddhist architecture, and by combining, retracting and releasing these spaces, the inner and outer courtyards and the main hall of the Huineng Memorial Hall are organically connected in a progressive way, creating an elegant and Zen-like space.

南京城墙博物馆改扩建工程

Expansion Project of Nanjing City Wall Museum

项目位置：江苏省南京市
设计时间：2017 年
竣工时间：未竣工
建筑规模：12,000 m^2

Location：Nanjing, Jiangsu
Design：2017
Completion：Under Construction
GFA：12,000 m^2

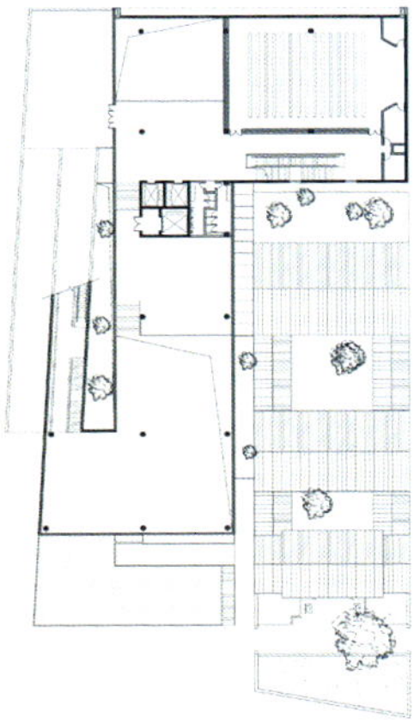

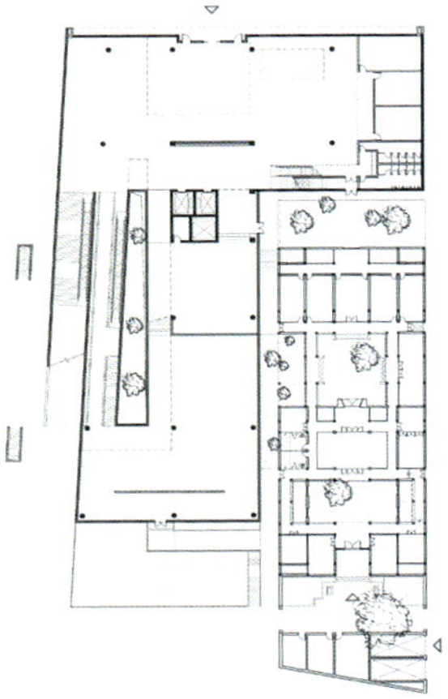

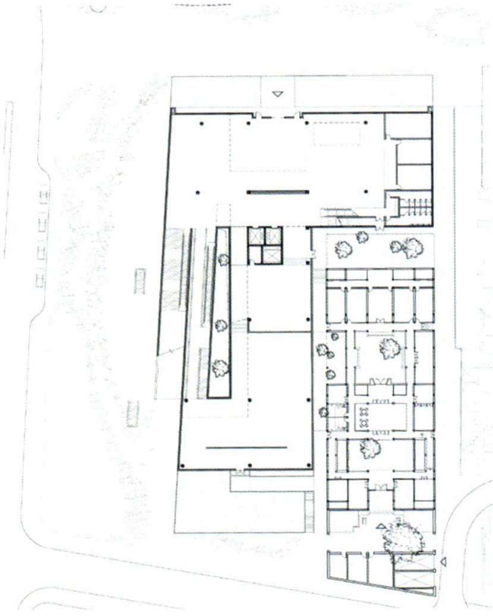

博物馆的形体设计试图将古城墙的马道、平台等元素加以提炼，形成近城低、远城高的自然态势，让博物馆真正地成为面向公众开放的城市公园，成为举办各种庆典活动的文化集散地。

南京城墙博物馆位于南京中轴极具历史价值的中华门以东。它不是一个普通的博物馆，它最重要的藏品不是在博物馆里，而是矗立在它面前的、宏伟的南京城墙。它应该是呈现城墙历史文化的装置，是融入城墙历史环境的艺术品，也是一个面向公众开放的城市公园。

The form of the design at the museum strives to refine elements of the ancient city wall, such as the bridleway and the platform, in order to produce a natural effect where such elements are low near the wall and higher with distance. This transforms the museum into a city park truly opens to the public and a cultural center for various celebration activities.

The Nanjing City Wall Museum lies east of the historically valuable Zhonghua Gate in the axis of Nanjing. This is not an ordinary museum, as its most important collection is not anything stored or exhibited at the museum but rather the magnificent city wall that stands before it. It should serve as a device of presenting the historical and cultural features of the city wall, a work of art that integrates the historical environment of the city wall, and a city park that is open to the public.

深圳市宝安中心区图书馆
Shenzhen Bao'an Centre Library

项目位置：深圳市宝安区
合作单位：严迅奇建筑师事务所
设计时间：2010 年
竣工时间：2013 年
建筑规模：48,000 m^2

Location：Baoan, Shenzhen
Partner：Rocco Design Architects Associates Ltd.
Design：2010
Completion：2013
GFA：48,000 m^2

深圳市宝安中心区图书馆位于深圳市宝安区文化核心区域的中央绿轴上，东临宝华路，西靠宝兴路，北以海秀路为界，南面向海滨开放。项目占地面积约 31,853 ㎡，总建筑面积 48,000 ㎡，其中图书馆面积 28,271 ㎡，博物馆建筑面积 4,045 ㎡。展览馆建筑面积 4,903 ㎡，地下车库及设备建筑面积 10,781 ㎡。建筑高 23.95m，地上有 4 层，其中二层设夹层；地库有 2 层，地下二层为车库。

Shenzhen Bao'an Centre Library is located on the central green axis of the cultural core area of Bao'an District in Shenzhen, facing Baohua Road on the east, Baoxing Road on the west, Haixiu Road on the north and the seaside on the south. The project covers an area of about 31,853 square meters, with a total construction area of 48,000 square meters, including 28,271 square meters for the library, 4,045 square meters for the museum, 4,903 square meters for the exhibition hall, 10,781 square meters for the underground garage and equipment, and 23.95 meters in height. There are four floors above ground, and the second floor has mezzanine. The basement has two floors, and the second floor is the garage.

厦门海峡旅游服务中心

Xiamen Strait Tourist Service Centre

项目位置：福建省厦门市
设计时间：2012 年
竣工时间：2019 年
建筑规模：387,300 m^2

Location：Xiamen, Fujian
Design：2012
Completion：2019
GFA：387,300 m^2

我们从用地的独特地理位置上找到灵感。海峡旅游服务中心选址地五通历史上就是通往宝岛台湾的咽喉。借用“一衣带水”这个深远的意义，方案采用了“适当离析”与“紧密联系”的两个概念重塑区块的建筑空间系统。“适当离析”是指将建筑单体分为两个组团，而围合形成带状的内部空间；“紧密联系”是指将建筑空间布局有意识地按功能特征聚拢布局。如此一来，均匀分布的地块空间通过建筑的有机聚散，形成极具动感的整体关系效果。而各个建筑物似分而连，在构成上保持了高度的统一，如同一脉相承。

We inspired from the unique geographical location of the land. In the history, Wutong where the tourist service center is located, has been the throat leading to Taiwan. Based on the far-reaching significance of “one strip of water”, the scheme adopts the two concepts of “appropriate segregation” and “close connection” to reshape the building space system of the block. “proper segregation” refers to the division of building monomers into two groups, and the formation of banded internal space. “close connection” refers to the conscious gathering of the layout of architectural space according to the functional characteristics. Therefore, the evenly distributed plot space is scattered through the organic accumulation of the building, thus forming a very dynamic overall relationship effect, and each building seems to be separated and connected, maintaining a high degree of unity in the composition, as if in one vein.

嘉善规划展览馆
Jiashan Urban Planning Exhibition Hall

项目位置：浙江省嘉善县
设计时间：2014 年
竣工时间：2016 年
建筑规模：10,000 m^2

Location：Jiashan, Zhejiang
Design：2014
Completion：2016
GFA：10,000 m^2

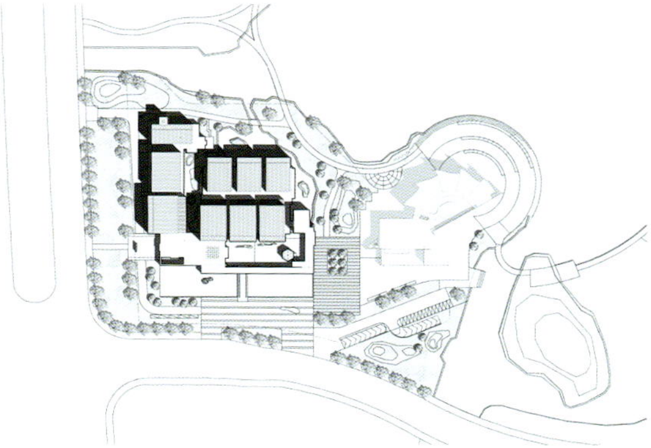

嘉善县地处江南，水赋予了它生长和发展的动力，积淀了深厚的文化内涵。设计概念“善水院”取嘉善西塘水乡与江南园林院落意象，将自然的水韵与人文的园意共融于建筑之中，以现代的手法演绎嘉善水乡底蕴，解读区域人文历史，展示全新城市形象，打造独具特色的吴越水乡新城的城市会客厅。

嘉善古韵，善水新院。嘉善规划展示馆的设计借鉴嘉善水乡地域性的建筑特色，整体结构处处与水为邻。建筑汲取了江南园林院落的精髓，构建以礼仪水庭、休闲景园与办公清院为主的一系列主题院落。

Located in the south of the lower reaches of the Yangtze River, water has always provided Jiashan County with the impetus for growth and development, and has thus acquired profound cultural connotations. For the design of the “Good Water Courtyard”, we selected the artistic concept of the Xitang waterside town in Jiashan and the Jiangnan garden courtyard. Then we integrated this with the natural rhythm of the water and the surrounding gardens. In this way, the cultural deposits of the Jiashan water town and its regional history have been interpreted in a modern way, creating a unique urban reception hall for this new water town.

The ancient charm of Jiashan, and the new courtyard of Shanshui (Good Water). The design of the Jiashan Urban Planning Exhibition Hall draws on regional architectural styles. The building derives its architectural essence from Jiangnan garden courtyards to create a series of spaces themed around the ceremonial water court, leisure gardens and quiet office compounds.

三祖文化博物馆

Three Ancestors（Sanzu）Cultural Museu

项目位置：河北省涿鹿县
设计时间：2013 年
竣工时间：2015 年
建筑规模：98,748 m^2

Location：Zhuolu, Hebei
Design：2013
Completion：2015
GFA：98,748 m^2

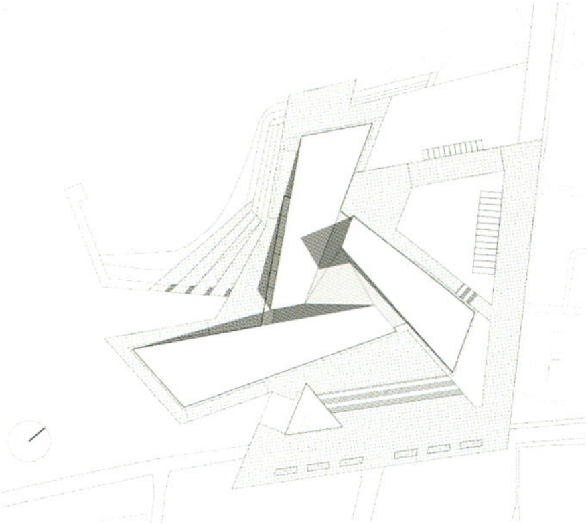

方案以三祖汇聚为出发点，结合地形将建筑分为三大体块，形成汇聚交融之势。以粗犷的巨石化作三祖文化的载体，建筑造型大气凝练，其寓意为涿鹿三祖文化之石。方案师法自然，因地制宜，通过抽象的几何造型，体现古文明的自然属性，整体景观呼应场地的山水格局。建筑三个体块围合形成时空交汇的文化体验场所。中庭空间取合符、融合之意，表达文明融合、中华一统的寓意。

Taking the starting point of the convergence of three ancestors, the museum is divided into three portions shape into a converging pattern. By imitating the raw physical form of natural stone, the museum is endeavored to create a powerful spirit of the place and echoes to the characteristics of the ancient civilization. The architectural scheme is ultimately generated from the nature building environment and derived from the local conditions. The new construction is reproducing the Chinese civilization through its neo-geometric shape, and definitely, echoing the whole landscape pattern. Three individual blocks of the new museum, forming a cultural experience construction of time and space. The central courtyard enclosed by three blocks symbolizes the harmony, vitality, as well as the unity.

天津博物馆
Tianjin Museum

项目位置：天津市
设计时间：2008 年
竣工时间：2012 年
建筑规模：55,000 m²

Location：Tianjin
Design：2008
Completion：2012
GFA：55,000 m²

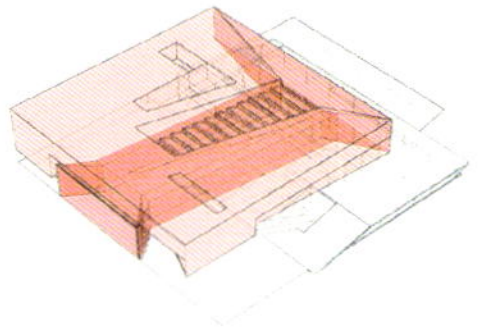

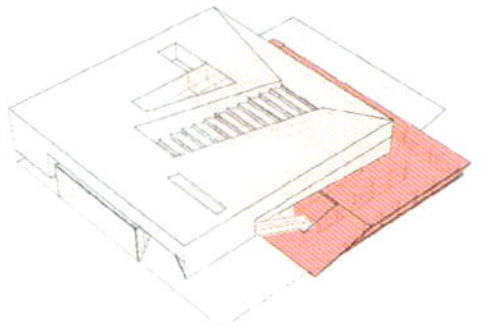

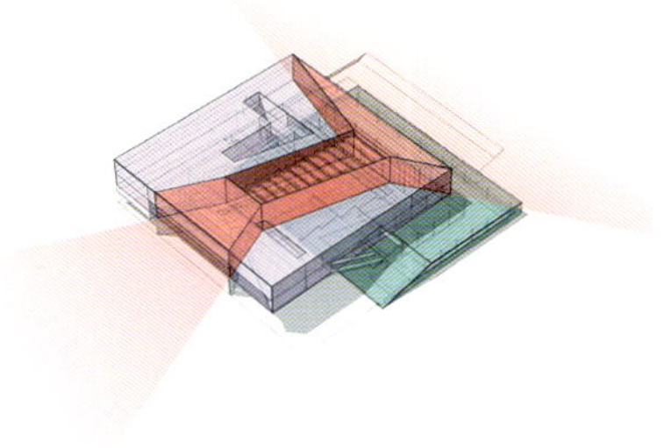

天津博物馆位于城市最核心的文化中心。我们以“时光隧道”作为主题，再现天津的悠久历史和深厚文化。它由内而外贯穿整个博物馆，宽敞宏大的公共大厅紧密联系各个展厅，成为公众参观和体验建筑的核心空间。空间的宽窄，光线的明暗，平台的高低产生连续而微妙的变化，给人以独特的感受。

Tianjin Museum is located in the Tianjin Cultural Center, a core area of the city. The “time tunnel” is the main theme reflecting the city’s long history and profound culture. It runs through the museum from the inside to the outside. The spacious public hall connects each exhibition area and becomes the focal point for visitors to experience the architecture. The size of the space, the bright lighting and the height of the platform produce continuous and subtle changes, giving people a unique experience.

嘉善规划展示馆

烟台文化中心

Yantai Cultural Centre

项目位置：山东省烟台市
设计时间：2007 年
竣工时间：2008 年
建筑规模：126,121 m^2

Location：Yantai, Shandong
Design：2007
Completion：2008
GFA：126,121 m^2

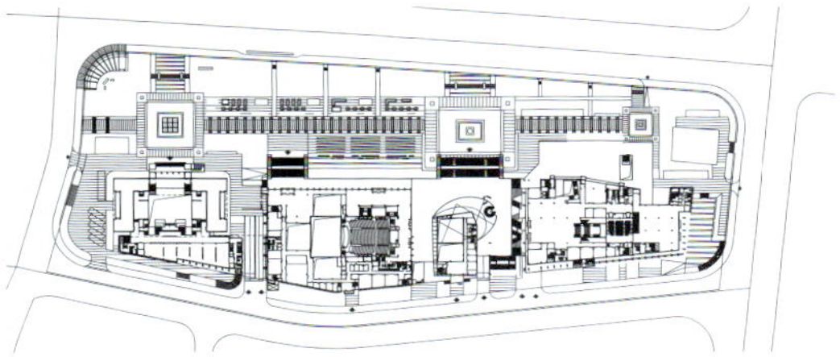

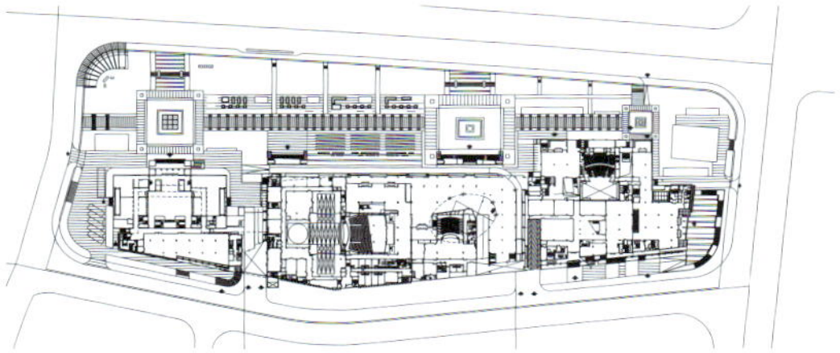

烟台文化中心是在城市中心紧凑的用地上规划建设的一座包含六大功能（大剧院、博物馆、群艺馆、京剧院以及青少年宫、书城）的十几万平方米的大型文化建筑，它体现了一种前瞻性的集约化土地使用策略，将六大功能体块通过平台、屋顶、连接体形成一个有机整体，以增加可识别性和标志性，达成形象鲜明、协调统一的效果。

创作的灵感来自一系列烟台的自然特色和物质文化特征，将抽象的文化符号通过建筑学的手段进行转化，确定了“山海仙境、开放烟台”的总体设计意向，并以“交汇之海、文化之石”为具体的形象特征。

Despite the dense land use in the center of the city, the Yantai Cultural Center is planned to be built there. It is a large-scale cultural facility spanning hundreds of thousands of square meters. The center has six key functions: theater, museum, public cultural center, Beijing Opera theater, youth palace and bookstore. Full use is made of the surface area, with the six functional elements integrated into an organic whole by means of the platform, the roof and the connecting body. This makes the building instantly identifiable, turning it into a vivid symbol of the area.

We drew inspiration from various aspects of the natural characteristics and features of Yantai, translated abstract cultural symbols by means of architecture, and finalized the overall design of a “fairyland boasting mountains and the sea”. In this way, “Yantai is opening to the world” thanks to the imagery of “the crossroads of the sea and the stone of culture”.

超豪华音乐剧《宝莱坞商人》门票热销中。
抢票热线6605566 6860885

梅州世界客商中心
The World Hakka Entrepreneurs Center

项目位置：广东省梅州市
设计时间：2015 年
竣工时间：2017 年
建筑规模：49,200 m^2

Location：Meizhou, Guangdong
Design：2015
Completion：2017
GFA：49,200 m^2

梅州世界客商中心位于有着“世界客都”之称的梅州市，是梅州市政府重点文化建筑项目。项目位于梅州市芹洋半岛城市新发展地段，西侧面朝梅江，与旧城区隔江相望。项目的定位为多功能的综合性文化中心，为两年一度的世界客商大会提供永久性的会址，为世界客商的文化交流、历史展示提供聚会的场所，为梅州市民提供用于文化展览、创客办公、休闲餐饮的公共文化设施。

The World Hakka Entrepreneurs Center is located in Meizhou, which is known as “Hakka capital of the world”. It is in the new urban development area of KiunYong Peninsula, the west side of which faces Meijiang River and the old city. The project is positioned as a multi-functional comprehensive cultural center, providing a permanent venue for the biennial World Hakka Business Conference, a place for cultural exchanges and historical exhibitions. In addition, it provides Meizhou citizens with public cultural facilities for cultural exhibitions, creator offices and casual catering.

世界客商中心
The World Hakka Entrepreneurs Centre
共享客都文明

增城城市馆

Zengcheng City Gallery

项目位置：广州市增城区
设计时间：2009 年
竣工时间：2015 年
建筑规模：43,090.1 m²

Location: Zengcheng, Guangzhou
Design: 2009
Completion: 2015
GFA: 43,090.1 m²

增城城市馆位于广州增城公共核心区域，基地东北毗邻荔枝文化公园，东面为增城图书馆。本项目是一个集科学技术馆、博物馆、档案馆、城建规划展览馆和人防 101 工程五大功能于一体的具有鲜明时代风格和岭南人文蕴意的科技文化中心，是增城重要的知识信息枢纽和精神文明建设基地。

增城城市馆以把握“历史、现代和未来”三个跨越时空的主题为设计主旨，运用矩形体块的立体组合效果和空间虚实搭配，营造出隐喻“智慧宝盒”的创作意向，巧妙地将四馆形象有机地统一起来。

Zengcheng City Gallery is located in the public core area of Guangzhou Zengcheng, adjacent to Litchi Culture Park in the northeast of the base, and Zengcheng Library in the east. This project is a science and technology center with distinct times style and Lingnan humanistic implication, which integrates the five functions of science and technology museum, archives, urban construction planning exhibition hall and Civil Air Defense 101 Project. It is an important knowledge and information hub and spiritual civilization construction base in Zengcheng.

Taking the theme of “history, modern and future” as the design gist, Zengcheng City Gallery utilizes the three-dimensional combination effect of rectangular body block and the collocation of space virtual reality to create the creative intention of metaphor “wisdom treasure box”, and skillfully unifies the image of the four pavilions organically.

广州古田村民中心
Guangzhou Gutian Villager Center

项目位置：广州市从化区
设计时间：2016—2017 年
竣工时间：2018 年
建筑规模：656 m²

Location：Conghua, Guangzhou
Design：2016-2017
Completion：2018
GFA：656 m²

村民中心位于村落中部的一块不规则场地上。为了给用地紧张的村落提供公共空间，建筑只占取场地东侧约 700 m²的用地。建筑的总平面呈“L”形，由一栋一层的坡顶会议厅和一栋三层的平顶综合楼组成，两者以廊道相接。建筑通过采用当地的材料和工艺，运用现代设计手法，因地制宜，对古田村的村落风貌、气候条件和村民的使用模式进行了积极的回应，是我国当代村民公共建筑的一次有益尝试。

The village center is located on an irregular site in the middle of the village. In order to provide public space to villages with tight use, the building only takes up about 700 square meters of land on the east side of the site. The general plane of the building is "L" shaped, consisting of a one-story sloping conference hall and a three-story flat-top complex, which are connected by a gallery. By adopting local materials and techniques, using modern design techniques and adapting to local conditions, the building responds positively to the village style, climatic conditions and villagers' use patterns of Gutian Village. It is a useful attempt for contemporary villagers' public buildings in China.

龙船山莞香文化公园二期——香馆

Longchuan Shanguan Incense Cultural Park Phase II – Incense Pavilion Project

项目位置：广东省东莞市
设计时间：2010 年
竣工时间：2014 年
建筑规模：7,500 m^2

Location：Dongguan, Guangdong
Design：2010
Completion：2014
GFA：7,500 m^2

香馆的设计具有内外两层表皮：内层是采用大面积的 low-e 玻璃幕墙作为盒子的内表皮，这是由“香文化”的展示具有的特殊性决定的，同时，大面积的玻璃面不仅使室内具有良好的采光，也有利于实现自然通风，减少对人工照明和通风的依赖，降低建筑的能耗；还有，通透的玻璃面消除了室内外空间的隔阂，充分利用公园的自然环境，使参观者在观看展品的同时，还能享受周边优美的湖光山色。香盒的外表皮是一层采用镂空手法的“虚”立面，一方面，虚的立面能够很好地表达香的主题，阐释禅的意境，尤其是晚上，配以灯光的营造，镂空的香盒内亮外暗，犹如一个透光的香炉，使香的主题得到升华；另一方面，虚的立面也有助于减少建筑的体量，使建筑更好地融入环境当中；还有，这层虚的外表皮还是一个很好的遮阳构件，有如岭南园林建筑中的花窗，有效地阻挡阳光的直射，体现了岭南建筑的传统特色。

Incense pavilion design is inside and outside two layer of the epidermis: the inner area of low-e glass curtain wall is used as the box inner epidermis, at this time due to the “Xiang” culture, there is a special show, at the same time, a large area of glass surface is not only to make indoor with good daylighting, but also to realize the natural ventilation, reduce our dependence on artificial lighting and ventilation, reduce building energy consumption. What's more, the transparent glass surface eliminates the gap between indoor and outdoor space, making full use of the natural environment of the park, allowing visitors to enjoy the beautiful lakes and mountains while watching the exhibits. The outer surface of the incense box is the hollow out facade. On one hand, the hollow facade can well express the theme of incense and explain the artistic conception of Zen. Especially at night, with the construction of lighting, the hollow incense box is bright inside and dark outside, is just like a transparent incense burner, which sublimates the theme of incense. On the other hand, the hollow facade also helps to reduce the volume of the building, making it better integrated into the environment. In addition, this hollow outer skin is also a good shading component, likes the flower window in Lingnan garden building, which effectively blocks direct sunlight and reflects the traditional characteristics of Lingnan building.

广州粤剧艺术博物馆
The Museum of Cantonese Opera Art

项目位置：广东省广州市
设计时间：2012 年
竣工时间：2016 年
建筑规模：17,436 m^2

Location: Guangzhou, Guangdong
Design: 2012
Completion: 2016
GFA: 17,436 m^2

粤剧艺术博物馆设计方案尊重基地与粤剧文化历史、岭南传统园林的密切关系，设计注重对粤剧艺术及岭南园林艺术精髓的深入挖掘，在设计中追求与粤剧艺术在精神内涵及园林建筑表现上的气韵相合，将建筑实体转化为园林空间，以求达成岭南两大文化标志跨越时空的当代对话。

The Museum of Cantonese Opera Art respects the close relationship between the site and the culture and history of Cantonese Opera and Lingnan traditional gardens. The design pays attention to the deep exploration of Cantonese Opera art and the essence of Lingnan garden art. Pursuing the harmony with the Cantonese Opera art in the spiritual connotation and the performance of landscape architecture, we transform the building entities into garden space in order to achieve a contemporary dialogue between the two symbols of Lingnan culture across time and space.

北滘市民活动中心

Beijiao Civic Activity Center

项目位置：广东省佛山市
设计时间：2014 年
竣工时间：2017 年
建筑规模：35,381 m^2

Location: Foshan, Guangdong
Design: 2014
Completion: 2017
GFA: 35,381 m^2

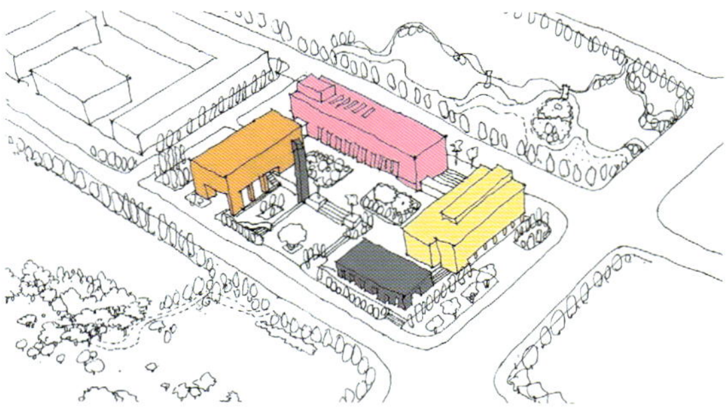

尊重新城区的总体规划，与城市周围空间环境协调融合，是本项目设计的指导思想。作为市民活动的主要场所，空间布局上采取开放、包容的态势，建筑沿南、北和西面布置，留出东南面广场与人民公园在空间上相衔接，成为面向文化公园的生活舞台。广场周围建筑物首层架空的灰空间透过通花砖墙，成为室内外空间连接的过渡和纽带，与岭南地区的地域和气候条件相适应。建筑立面设计理性而不刻板、创新而不铺张，避免高档材料的堆砌以及张扬的商业形式，还市民一个开放、平和、人性的休憩活动场所，与周围的城市广场、公园、文化中心等共同构成一个系统化的、网络化的、人性化的城市空间。

Respect for the overall planning of the new urban area and coordination and integration with the urban surrounding space environment are the guiding ideology of this project design. As the main venue for civic activities, the layout of the space is open and inclusive. The buildings are arranged along the south, north and west. The southeast plaza and the Beijiao People's Park are connected in space, making it a life stage for cultural parks.The grey space of the first floor of the building around the square passes through the brick wall, which becomes the transition and link between the indoor and outdoor space, and is compatible with the geographical and climatic conditions of Lingnan. The design of the building facade is rational but not rigid, innovative and not extravagant, avoiding the stacking of high-grade materials and publicized commercial forms, and also providing citizens with an open, peaceful and humane recreational activity place, together with the surrounding city squares, parks, cultural centers and other forms of a systematic, networked and humanize urban space.

广州国际残障人文化交流中心

Guangzhou International Disability Cultural Exchange Center

项目位置：广东省广州市
设计时间：2007 年
竣工时间：2012 年
建筑规模：13,002 m^2

Location: Guangzhou, Guangdong
Design: 2007
Completion: 2012
GFA: 13,002 m^2

建筑隐喻“希望与关爱之花”，唤起人们对残障人士的关爱和对生活的美好希望。设计由自然形态中的花、叶子等有效的功能形式启发而来，不仅在形式上，也在功能上为残障人士使用提供了极大便利。

项目用地位于整个基地的南侧，在两条城市道路的转角处，可用范围大致呈东西走向。机动车出入口设于用地西面。7m 宽的车行道路环通整个场地，兼作消防车道。建筑物主入口的雨篷，提供了充分的全天候停车与上下客空间。地下车库出入口坡道设于用地北侧，为了方便下肢残障人士，在主体建筑的室外北面设置一排轮椅用机动车位，并设置专用标识。在本项目的基地内，按《方便残疾人使用的城市道路和建筑物设计规范》的规定，对所有的人行道、建筑物出入口均做了无障碍设计。

The architectural metaphor “flowers of hope and care” evokes people’s love for the disabled and good hope for life. Inspired by effective functional forms such as flowers and leaves in natural form, it is greatly facilitated not only in form but also in function for the use of disabled people.

The project site is located on the south side of the entire base. At the corners of the two urban roads, the available range is roughly east-west. The motor vehicle entrance and exit is located on the west side of the site. The 7-meter-wide roadway runs through the entire site and doubles as a fire lane. The awning of the main entrance of the building provides full 24×7 parking and space for guests. The underground garage entrance and exit ramp is located on the north side of the site. In order to facilitate the disabled of the lower limbs, a row of wheelchair parking spaces is set up on the north side of the main building, and a special sign is set. In the base of the project, all walkways, building entrances and exits. have been designed to be barrier-free according to the “Code for Design of Urban Roads and Buildings for the Disabled”.

西关美食文化博物馆

Xiguan Food Culture Museum

项目位置：广东省广州市
设计时间：2013 年
竣工时间：2018 年
建筑规模：4,410.6 m^2

Location：Guangzhou, Guangdong
Design：2013
Completion：2018
GFA：4,410.6 m^2

设计深入调研了荔枝湾的历史，关注场地的现状，思考场地在整个区域中所扮演的角色，确立了以建筑与历史对话为主旨的总体设计定位，将现代化的停车场功能置入地下，地面上着力研究和吸收岭南庭园和西关传统建筑的特质，以现代的设计思维，精心设计，大胆构思，形成了可游可居的园林酒家，创造蕴含传统气质又展示现代气息的新西关建筑群。

The design thoroughly investigates the history of Lichee Bay, pays attention to the current situation of the venue, considers the role of the venue in the entire region, establishes the overall design orientation of the architectural and historical dialogue, and puts the modern parking lot function into the ground and ground. Focusing on the research and absorption of the characteristics of Lingnan garden and Xiguan traditional architecture, with modern design thinking, careful design, bold ideas, the formation of a gardenable wine house, can create a new Xiguan complex with traditional temperament and modern atmosphere.

南海会馆
Nanhai Guild Hall

项目位置：广东省佛山市
设计时间：2011—2012 年
竣工时间：2021 年
建筑规模：176,259 m^2

Location：Foshan, Guangdong
Design：2011–2012
Completion：2021
GFA：176,259 m^2

南海会馆位于西樵山北麓，听音湖之滨，定位为岭南地区的“精神祠堂、文化殿堂”。南海会馆由南海会堂、飞鸿馆和有为馆共同构建。三大建筑围绕听音湖形成品字形格局，突出会堂的主体建筑形象，与山水景观、场地环境相和谐。

南海会馆试图在今天唤起历史传统的纵深感，在材质与工艺的现代转译中，我们遵循品相对等原则——在现代设计抽象和简化过程中，注重保留传统特色构成、形态、色彩、质感等要素在知觉上的同一性，落实为现代技术支持下的具体形式。

Nanhai Guild Hall is located in the north foot of Xiqiao Mountain, on the shore of Tingyin Lake. It is positioned as the “spiritual ancestral hall and cultural temple” in Lingnan area. Nanhai Guild Hall is constructed by Nanhai Hall, Feihong Hall and Youwei Pavilion. It forms a triangle pattern around Tingyin Lake to highlight the main architectural image of the hall, which is in harmony with landscape and site environment.

Nanhai Guild Hall tries to arouse the depth of the historical tradition. In the modern translation of materials and craftsmanship, we follow the principle of relative product—in the process of abstraction and simplification of modern design, we pay attention to retaining the perceptual identity of traditional characteristic composition, form, color, texture and so on, and implement it as a concrete form supported by modern technology.

迪拜清真寺

Scheme Design of Dubai Mosque

项目位置：阿拉伯联合酋长国，迪拜
设计时间：2016 年
竣工时间：未竣工
建筑规模：25,225 m^2

Location：Dubai, The United Arab Emirates
Design：2016
Completion：Under Construction
GFA：25,225 m^2

我们通过采用适当的建筑策略，将传统的伊斯兰文化与现代建筑词汇相结合，努力将项目打造成具有独特形式的代表性建筑，以至成为当地甚至迪拜的地标。此外，我们致力于设计一个可以展示现代城市的生态性和公共性的建筑。建成后，清真寺将超越纯粹的建筑空间，为整个城市空间和景观绿轴增添光彩。籍此，清真寺及其周围的绿轴将成为城市公共活动的重要场所。这里的文化活动具有伊斯兰特色，公开、公平、有仪式感。该项目的亮点在于它的亲和可达性，它将在这里与每个人分享宗教的精神。此外，伊斯兰文化的融入进一步提升了建筑的文化价值。

By adopting proper building strategy that combines the traditional Islam culture with modern architectural vocabulary, we endeavor to develop the Project into a representative building with unique form, which will inevitably become the icon of the region and even Dubai. In addition, we are committed to designing a cultural landmark which can showcase the modern, ecological and public character of a city. Once completed, the mosque will go beyond a pure architectural space and add brilliance to the whole urban space and landscape green axis. Consequently, the mosque and its surrounding green axis become an important place for urban pubic activity. Characterized by Islam features, the cultural activities here will take place in an open, fair and ceremonial manner. The highlight of the project lies in its accessibility where it shares the religious spirit with everyone. In addition, the incorporation of the Islam culture further enhances the cultural value of the building.

花山岩画游客服务中心
Huashan Rock Visitor Center

项目位置：广西壮族自治区崇左市
设计时间：2018 年
竣工时间：2021 年
建筑规模：14,402 m^2

Location: Chongzuo, Guangxi
Design: 2018
Completion: 2021
GFA: 14,402 m^2

花山岩画游客中心项目选址在崇左市宁明县花山镇木州村花山岩画文化风景区，建设内容为游客中心，主要涵盖陈列厅、报告厅等教育展览功能及售票、餐饮、购物等配套服务功能。整体设计以体现广西的钟灵毓秀、花山岩画的独特壮观为出发点，利用自然起伏的体量和大架空的空间，使游客中心的建筑形体融入周围环境，塑造出既有花山景区特色又亲切宜人的游览环境。

Huashan Rock Visitor Center is in Huashan Rock Cultural Tourist Attraction, Muzhou Village, Huashan Town, Ningming County, Chongzuo City. It is a visitor center containing education and exhibition functions including exhibition hall and lecture hall, as well as supporting service functions including ticketing, catering, and shopping. The overall design reflects the unique views and Huashan Rock paintings in Guangxi as the starting point. And it makes use of the natural undulating dimension and large overhead space so that the architectural form of the visitor center integrates into the surroundings, shaping both the characteristics and the friendly environment of Huashan Tourist Attraction.

崇左市文化艺术中心

Chongzuo Culture and Art Center

项目位置：广西壮族自治区崇左市
设计时间：2011 年
竣工时间：2021 年
建筑规模：13,669 m^2

Location：Chongzuo, Guangxi
Design：2011
Completion：2021
GFA：13,669 m^2

设计以“城市中的景观、景观中的建筑”作为设计切入点，从壮族地域文化中提炼出“锦绣梯田”的设计理念，与建筑的造型与空间营造巧妙结合，注重建筑的地域、文化、民族的特性，用现代手段演绎原汁原味的壮族文化。

注重整体协调，建立“湖面—建筑—广场”三位一体的整体格局，采用掏空、梯田跌落等手法进行切割；将公众娱乐性掺入设计之中，将建筑内部灰空间、沿湖景观空间甚至屋顶均设计为开放性空间面向市民；将地域性、民族性特点要素加入建筑之中，使文化艺术中心具有区域标识性。方形的建筑体量与圆形天穹空间来源于中国传统“天圆地方”的精神文化，在立面处理上采用现代技术处理，展现出建筑的时代性。

The design is based on the “landscape in the city, the architecture in the landscape” as the design entry point. The design concept of “Jinxiu terraces” is extracted from the regional culture of Zhuang nationality, which is ingeniously combined with the modeling and space of the building, paying attention to the geographical , cultural and national aspects of the building. The characteristics of the nation, using modern means to interpret the original Zhuang culture.

The design of the center pays attention to the overall coordination, establishes the overall pattern of the “lake-architecture-square” trinity, adopts the method of hollowing out, terraced fall and other methods; puts the public entertainment into the design, the interior gray space of the building, the landscape space along the lake and even the roof are all designed to open-ended space for the public; to incorporate regional and national characteristics into the building, so that the cultural and art center has regional identity. The square building volume and the circular sky space are derived from the spiritual culture of the traditional Chinese “round sky and square earth”, which is treated with modern technology in the façade , showing the modernness of architecture.

镇江市图书馆
Zhenjiang Library

项目位置：江苏省镇江市
设计时间：2014 年
竣工时间：2017 年
建筑规模：63,269 m^2

Location: Zhenjiang, Jiangsu
Design: 2014
Completion: 2017
GFA: 63,269m^2

项目从图书馆的文化属性出发，结合当地的航运文化、山水文化、文学文化、吴文化，创造性地提出“西津泛舟、北固登峰，文宗藏卷、镇图汇萃”的设计理念。

图书馆通过结合中央绿轴的景观设计营造整体的书香公园的文化气氛，通过整个二层架空营造一个开放的城市客厅空间，通过变化的立面肌理营造图书馆建筑的文化气质，通过首层平层进入的方式营造和谐亲民的入口形象，通过大台阶中庭阅读空间的设置营造一个书山书海的震撼形象，通过东侧开放式的多层退台流通部营造一个内外皆景的多层次阅读氛围，通过图书馆和文创商业有机结合的方式营造一个相辅相成、充满活力的新型文化综合体空间。

The project starts from the cultural attributes of the library, combines the local shipping culture, landscape culture, literature culture, Wu culture, and creatively proposes the design concept of “Xijin boating, Beigu Climbing, Wenzong Tibetan scroll, town map collection”.

The library creates a cultural atmosphere of the overall Shuxiang Park through the combination of the central green axis landscape design, creating an open urban living room space through the entire two-story overhead, and creating the cultural temperament of the library building through the changing facade texture, through the way of entering the first floor leveling creates a harmonious and intimate entrance image, and creates a shocking image of numerous books through the setting of the large atrium reading space, and creates a multi-story reading atmosphere inside and outside through the open-ended multi-level retreat circulation department on the east side. Through the organic combination of the library and the cultural and creative business, the library creates a new and comprehensive cultural complex space that complements each other.

丽江市博物院新馆
Lijiang Museum

项目位置：云南省丽江市
设计时间：2015 年
竣工时间：2018 年
建筑规模：13,428 m^2

Location: Lijiang, Yunnan
Design: 2015
Completion: 2018
GFA: 13,428 m^2

丽江博物院位于丽江市大研古城到玉龙雪山的必经之路上，在束河古镇以东，白沙古镇东南方向。作为丽江少数的大体量公共文化建筑，设计力求在与古城环境尺度协调的基础上保持公共建筑应有的张力。

以“山水之灵、瓦屋栉比”为设计概念，表达瓦屋鳞次的古城印象、人文栉比的精深文化，并融入丽江的山川地理、历史人文的抽象写意，展现雪山潭泉的山水之灵、古道重镇的演进历史、多元文化的古朴之风、舒缓休闲的柔软之地等独特的自然人文精粹。

The Lijiang Museum is located on the only road from Dayan Ancient Town to the Yulong Snow Mountain, east of Shuhe Ancient Town and southeast of Baisha Ancient Town. As it is one of the rare public cultural buildings in Lijiang, we strove to maintain the style of public buildings while remaining true to the environment of the ancient city.

The design was based around the concept of the “spirit of the mountains and waters, closely aligned to the tiled-roof houses”. We tried to express the impression of closely arranged tiled—roof houses and the profound culture of the ancient city, and to integrate the geography of the mountains and rivers with an abstract historical and cultural snapshot, so as to exhibit the unique essence of the area. This includes the spirit of the snow mountain and the pool spring, the history of this important town on the ancient trading routes, the primitive style of multiculturalism and the serene and tender land.

02 办公建筑
OFFICE ARCHITECTURE

珠江新城西塔	International Financial Center of Guangzhou Zhujiang New Town（West Tower）
广州利通广场	Guangzhou Leatop Plaza
广州越秀金融大厦	Guangzhou Yuexiu Financial Building
广州天伦控股大厦	Guangzhou Talent Holdings Building
广州长隆大厦	Chimelong Tower (Guangzhou)
日立电梯（中国）有限公司研发楼	Hitachi Elevator (China) Co., Ltd. R&D Building
东莞生态园控股有限公司办公楼工程设计	Office Building of Dongguan Eco—park Holding Co., Ltd. Project Design
柳州九洲大厦	Liuzhou Jiuzhou Building
南宁九洲国际大厦	Nanning Jiuzhou International Building
广州南站核心区 BA01077 地块设计（路福联合广场）	Block BA01077 (Rofo Union Plaza)，Core Area of GuangZhou South Railway Station
武汉光电创新园（一期）	Wuhan Optoelectronic Innovation Park (Phase I)
东莞松山湖科技产业园区科学苑	Science Park, Gongguan Songshan Lake Science and Technology Industrial Park
渔安·安井温泉旅游城“未来方舟”B5、B6 组团	Yu'an Anjing Hot Spring Tourism Town “Future Ark” B5 and B6 Groups
仁恒滨海中心	Yanlord Marina Center
中国体育彩票广西营销展示中心	China Sports Lottery Guangxi Marketing Exhibition Center
君豪国际商业城	Junhao International Business Town
西樵纺织科技大厦	Xiqiao Textile Technology Building
广发证券大厦	Guangfa Securities Headquarters
深圳后海天佶湾国际广场项目	Shenzhen Houhai Tianji Bay International Plaza

珠江新城西塔

International Financial Center of Guangzhou Zhujiang New Town（West Tower）

项目位置：广东省广州市
合作单位：英国威尔金森埃里建筑师事务所；
奥雅纳工程顾问有限公司
设计时间：2004—2005 年
竣工时间：2010 年
建筑规模：453,865 m^2
建筑高度：434 m

Location：Guangzhou, Guangdong
Partner：Wilkinson Eyre Architects，ARUP
Design：2004–2005
Completion：2010
GFA：453,865 m^2
BH: 434 m

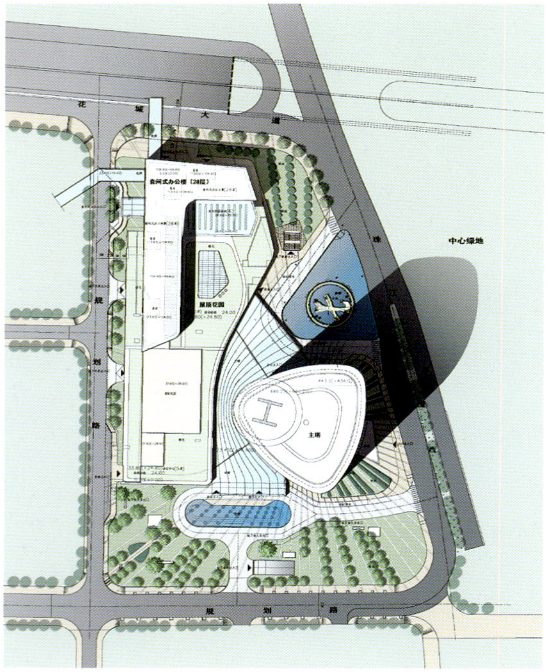

IFC
广州农村商业银行
富力
合景泰富

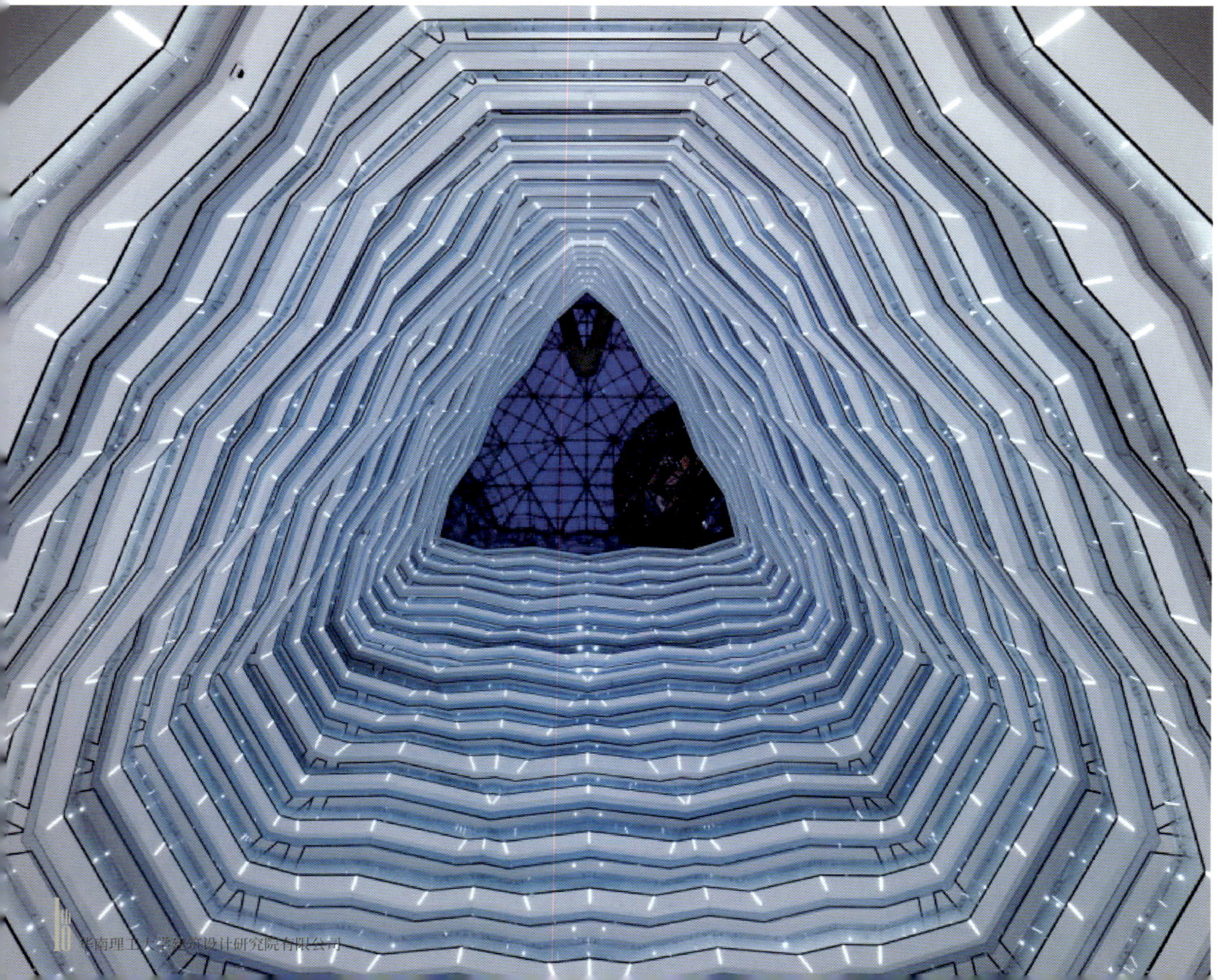

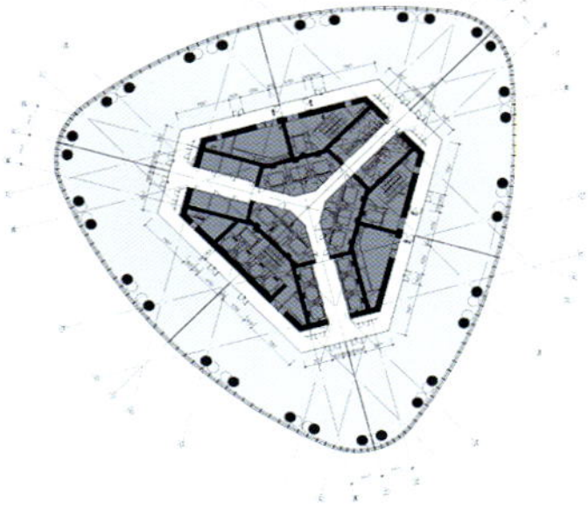

广州珠江新城西塔位于珠江新城核心商务区的中轴线西侧，连接了北面的商业贸易中心，南面为广州少年宫。

珠江新城西塔的标准层平面采用圆弧三角形，最大限度地保证了使用者能享受珠江及中轴线绿化区的美景。以曲线形状及透明的光滑建筑立面为主要思路设计，通过渐变宽度形成两头小中间大，在高度的三分之一处达到最宽，显露出建筑物的纺锤外形。建筑结构采用巨型斜交网格支撑体系，呈现出流动几何图案，并和楼板的分割及幕墙体系产生视觉对比。“钻石形状斜网格”结构随着塔的高度，逐步进行尺寸缩减。这种延伸性的设计使楼体呈现“直入云霄”的视觉效果。

珠江新城西塔耗资约 60 亿元，耗时 5 年建成，2010 年亚运会时期投入使用。塔楼总高度 434m（不含直升机平台高度），共 103 层，由智能化超甲级写字楼、白金五星级酒店及裙楼配套区域、套间式办公楼、超高档商场组成。该项目曾获英国皇家建筑协会国际奖、2013 年中国优秀工程勘察设计奖（建筑工程公建）一等奖。

The International Financial Center of Guangzhou Zhujiang New Town(West Tower) is located in the core business district of Zhujiang New Town, on the west side of the central axis, connecting the commercial district the north and the Pearl River in the south.

The standard floor of The International Financial Center of Guangzhou Zhujiang New Town(West Tower) adopts a shape of arc triangle, which ensures the user can enjoy the beauty of the Pearl River and the green area in the central axis. Designed with a curved shape and a transparent smooth building facade as the main idea, the top and the bottom of the architecture is small while the middle is large by the gradual width. And it reaches the widest at one-third of the height, revealing the spindle shape of the building. The building structure adopts a giant oblique grid support system, which presents a flow geometric pattern and visual contrast with the division of the floor and the curtain wall system. The “diamond-shaped diagonal grid” structure scales down as the tower heights gradually. This extensible design gives the building a visual effect of “straight into the sky”.

The International Financial Center of Guangzhou Zhujiang New Town(West Tower) cost about 6 billion yuan, which took five years to complete and was put into use during the 2010 Asian Games. The total height of the tower is 434 meters (excluding the height of the helicopter platform), a total of 103 floors. It includes intelligent super-grade office buildings, platinum five-star hotels and podium supporting areas, suite office buildings, ultra-high-end shopping malls. The project won the Royal Society of Architecture International Award and was awarded the first prize in the 2013 Excellent Engineering Survey Design Award (Construction Engineering Public Building).

广州利通广场
Guangzhou Leatop Plaza

项目位置：广东省广州市
合作单位：墨菲 / 扬建筑师事务所
设计时间：2007 年
竣工时间：2010 年
建筑规模：159,717 m^2
建筑高度：302.7m

Location：Guangzhou, Guangdong
Partner：Murphy / Jahn Architects
Design：2007
Completion：2010
GFA：159,717m^2
BH: 302.7m

广州利通广场是珠江新城城市中轴线北大门收口上的一个关键项目。项目用地面积为 9,915.65 ㎡，建筑面积约 159,717 ㎡，塔楼总高度 302.7m，其中地上建筑总层数为 58 层，地下总层数为 5 层。地上部分主要包括超甲级写字楼、会议厅，地下部分主要包括设备用房、地下车库、装卸货区和少量的餐饮设施。建筑主体简洁、高雅、线条清晰、具有精致的几何造型。通过特殊的幕墙设计，修长的建筑形体被分成三个部分：底部是数层楼高的大堂，中间主体部分是办公楼层，顶部则是一个空中大厅。空中大厅具有一个鲜明的斜坡屋面，层层玻璃营造出雕塑的效果，就像是一块晶莹剔透的冰晶体。塔楼的设计轻盈、通透，并充分考虑了高效与节能。它将观景视角最大化，强调了与城市的连接及二者之间的关系。这座具有柔和动感的高楼挑战着我们看待超高层建筑的常规方式。整个项目结构设计合理，技术先进。塔楼采用“斜撑框架＋混凝土核心筒”的结构体系，具有施工便捷、造价低廉、工期较短的特点。

Guangzhou Leatop Plaza is a key project in the north of the central axis of Zhujiang New Town. The land area of the project is 9,195 square meters, the construction area is about 159,800 square meters, and the total height is 302.7 meters. The main functions of the 58-story tower include offices and conference halls. Garage, equipment rooms, loading areas and a small number of dining facilities are set in the 5-story underground part. The main body of the building is simple, elegant, with clear lines and exquisite geometric shapes. Through a special curtain wall design, the slender architectural form is divided into three sections: lobby in the bottom, office in the centre and a sky hall in the top. The sky hall has a distinctive sloping roof, and the layered glass creates a sculptural effect, like a crystal clear ice crystal. The design of the tower is light and transparent, with full consideration of efficiency and energy saving. It maximizes the viewing angle and emphasizes the connection to the city and its relationship. This soft and dynamic high-rise challenges the normal way we look at super-tall buildings. The entire project structure is reasonable in design and advanced in technology. The tower building adopts the straightening frame and concrete core tube architecture, which has the characteristics of convenient construction, low cost and short construction period.

广州越秀金融大厦
Guangzhou Yuexiu Financial Building

项目位置：广东省广州市
设计时间：2010 年
竣工时间：2013 年
建筑规模：210,360 m²
建筑高度：295.5 m

Location: Guangzhou, Guangdong
Design: 2010
Completion: 2013
GFA: 210,360 m²
BH: 295.5m

本项目位于珠江新城，广州新城市中轴线的核心区域，总用地面积 10,836 ㎡，总建筑面积 210,360 ㎡，共 68 层，建筑高度 295.5m。地上部分主要包括甲级写字楼会议室、少量辅助性的商业设施及公共服务配套设施。地下部分为车库、设备用房和人防设施。本项目采用带加强层的“斜撑框架＋混凝土核心筒”的结构体系。建筑具有以下特点：

环境友好的建筑。建筑总平面逆时针偏转约 30 设置，使写字楼的南向景观面最大化，同时与西边的珠江城大厦形成十分协调的呼应关系，顺应城市肌理，与未来城市中轴线形成良好的对应关系，成为整个城市中轴空间的重要组成部分。

造型创新的建筑。建筑从梭形的中央绿带得到灵感，以优雅的曲线和动感的折线形成修长、简洁且具有精致几何造型的塔楼，既似萌发的春芽，又似虔诚的合十双掌，摆脱了呆板的传统方盒子写字楼形象，形成地标。

绿色高效的建筑。建筑采用多种绿色建筑技术，尤其是带水平遮阳的幕墙系统，使本幢大厦成为节能环保的先锋。

The project is located in Zhujiang New Town, the core area of the central axis of Guangzhou New City. The total land area is 10,836 square meters, the total construction area is 210,360 square meters, a total of 68 floors, 295.5 meters in height. The main functions of the building include offices, meeting rooms, a small number of auxiliary commercial facilities and public service facilities. The underground part is for garages, equipment and civil air defense facilities. The structure of the project adopts the straightening frame and concrete core tube architecture. The building has the following characteristics:

Environmentally friendly building. The total plane of the building is rotated by about 30 counterclockwise to maximize the southward landscape of the office building. At the same time, it has a very harmonious relationship with the Pearl River Tower in the west. It conforms to the urban texture forms a good correspondence with the future urban central axis, and becomes an important part of the axis space throughout the city.

Innovative building. Inspired by the fusiform central green belt, the building forms a slender, simple and elegant geometrical tower with elegant curved lines and dynamic fold lines. It is like a spring bud and a devout pair of palms. The shape design, gets rid of the rigid traditional square box office building image, forming a landmark.

Green and efficient building. The building uses a variety of green building technologies, especially the curtain wall system with horizontal sunshade, making this building a pioneer in energy conservation and environmental protection.

創興銀行

广州证券

广州天伦控股大厦

Guangzhou Talent Holdings Building

项目位置：广东省广州市
设计时间：2010—2014 年
竣工时间：2015 年
建筑规模：46,999 m^2
建筑高度：99.9 m

Location：Guangzhou, Guangdong
Design：2010–2014
Completion：2015
GFA：46,999 m^2
BH: 99.9m

广州天伦控股大厦位于广州市城市密度最高的天河老城区，周边建筑风貌多样，北邻火车铁路，南侧为"城中村"林和村，传统的村落民宅与新建的高层住宅等共同构成其独特的城市环境。该方案在一开始即着重解决尺度及密度所带来的影响，通过体块的错动、开窗的形式及密度上的变化，达到消隐建筑体量、降低建筑尺度的目的，从而达到与周围街区风貌的和谐。

为呼应地形和规划退缩要求，建筑的裙楼被设计成不规则多边形，塔楼则为平行四边形，在相对合理的形状下争取使用面积最大化。建筑体量和造型元素也沿用了平面多边形和斜线切割等处理手法，强调了设计手法的一致性。此外，多单元、模块化的穿孔金属板表皮设计使建筑具备了低调且独特的外观。

Guangzhou Tianlun Holdings Building is located in Tianhe District, and old district with the highest density in Guangzhou. The surrounding buildings are diverse in style. The railway is in the north, and the residential area is in the south. The traditional houses and newly built high-rise houses form a unique urban environment. At the beginning, the program focuses on addressing the effects of scale and density. Through several adjustments in the volume and facade, the purpose of making building being harmony with the surrounding neighborhoods is achieved.

In order to meet the terrain and government requirements, the building's podium is designed as an irregular polygon, and the tower as a parallelogram to maximize the use of the area in a relatively reasonable shape. Building dimension and styling elements also follow the treatment of plane polygons and slash cuts, emphasizing the consistency of design techniques. In addition, the use of modular perforated sheet metal skin gives the building a low profile but unique outlook.

广州长隆大厦
Guangzhou Chimelong Tower

项目位置：广东省广州市
设计时间：2017 年
竣工时间：2018 年
建筑规模：104,422 m^2

Location：Guangzhou, Guangdong
Design：2017
Completion：2018
GFA：104,422 m^2

长隆大厦位于广州长隆度假区内，基地周围绿树成荫，环境宜人。设计构思根据建筑所在的场所环境，取岭南榕树之形，表绿色生态之意，创人性化之境。绿植由裙楼屋顶花园沿塔楼形体顺势而上，形成多层次的立体园林空间，打造舒适的办公微环境，创造一处独具特色的人性化办公建筑。

我们希望通过将形式、空间、功能、景观相结合，打造一栋具有时代感的生态、绿色建筑，创建一扇人性化的欢乐视窗。人们可在此愉悦办公，悠闲休憩，最终实现办公与休憩的双重功能，建筑与生态共融。

Embraced by a forest of banyans with verdant canopies, the Chimelong Tower in Guangzhou Chimelong Tourist Resort enjoys an attractive environment. The design concept, which is inspired by the project context, references the shape of banyan tree to create an ecological and human-friendly environment. Green plants grow from the roof garden of the podium to the tower body, creating a multi-level garden space which in turn helps to create a pleasant micro office environment and a human-friendly office building with unique features.

The design's goal is, by integrating the building form, space, function and landscape, to create an ecological green building with the feature of time, making it a pleasant and human-oriented showcase where people enjoy both work and leisure which can eventually realize the integration of business and leisure, and the harmony between the building and ecology.

日立电梯（中国）有限公司研发

Hitachi Elevator (China) Co., Ltd. R&D Building

项目位置：广东省广州市
设计时间：2008—2009 年
竣工时间：2014 年
建筑规模：23,874 m²

Location: Guangzhou, Guangdong
Design: 2008–2009
Completion: 2014
GFA: 23,874 m²

日立电梯（中国）有限公司研发楼位于广东省广州市番禺区大石镇石北工业区，占地面积约 4,000 m²，建筑面积 23,874 m²，建筑高度 27m，6 层，总投资约 1.2 亿元，是日立电梯中国区域的总研发中心。其选址位于厂区中轴，南面正对厂区入口草坪广场，建筑分南北两体量，南北长 152m，东西长 92m。建筑内部布局规整，外立面采用清水混凝土和 low-e 玻璃，虚实对比强烈，造型简洁有力。

日立电梯（中国）有限公司研发楼是我院近年来涉外研发建筑的一个优秀作品，总体布局顺应厂区整体规划的原则，在建筑物前方留出广阔的草坪广场，营造出良好的整体空间。建筑平面布局充分考虑了企业的运作模式，满足了企业办公的需求并提升了企业的研发效率。造型上运用了清水混凝土的技术，结合清水玻璃，形成虚实对比，塑造了高雅、简洁、大方的效果，赋予建筑科技感，提升了企业形象。

The R&D building of Hitachi elevator (China) Co., Ltd. is located in Shibei Industrial Zone, Dashi Town, Panyu District, Guangzhou City, Guangdong Province. It covers an area of about 4000 square meters and a building area of 23,874 square meters. The building is 27 meters high with six floors and a total investment of about 120 million yuan. The site is located on the axis of the factory, facing the lawn square at the entrance of the factory in the south. The building is divided into two dimensions, 152 meters long from north to south and 92 meters long from east to west. The internal layout of the building is neat, and the exterior facade is made of fair-faced concrete and low-e glass, with a strong contrast between the real and the virtual, and a simple and powerful shape.

The R&D building of Hitachi elevator (China) Co., Ltd. is an excellent work of our institute's foreign-related R&D building in recent years. The layout of the building fully considers the operation mode of the enterprise, meets the requirements of the enterprise office and improves the efficiency of the enterprise's research and development. The modeling uses the technology of fair-faced concrete, combined with fair-faced glass, to form a virtual and real contrast, shaping an elegant, concise, generous effect, endowing architecture with a sense of science and technology, and improving the corporate image.

柳州九洲大厦

Liuzhou Jiuzhou Building

项目位置：广西壮族自治区柳州市
设计时间：2010 年
竣工时间：2014 年
建筑规模：103,515.45 m^2

Location: Liuzhou, Guangxi
Design: 2010
Completion: 2014
GFA: 103,515.45 m^2

柳州九洲大厦规划总用地面积 21,386 ㎡，总建筑面积 103,515.45 ㎡，地上最高 35 层，地下 2 层，建筑高度 162.9m；为集办公、酒店、商业为一体的综合建筑，定位为柳州市柳江边的地标建筑。

柳州市是山水城市，青山环绕，水抱城流，“城在山水园林中，山水园林在城中”，我们把“龙游山水”概念融入建筑造型，将错落的建筑体量、丰富的立面虚实对比，在建筑立面上形成连续的折线图案，抽象构造游龙的形象，营造典雅明快的建筑造型，形成柳江边独树一帜的城市名片。

Liuzhou Jiuzhou Building has a total planned land area of 21,386 square meters, a total construction area of 103,515.45 square meters, a maximum of 35 floors above ground, 2 underground floors and a building height of 162.9 meters. It is a comprehensive building for office, hotel and business. It is positioned as a landmark building on the Liujiang River in Liuzhou City.

Liuzhou City is a mountainous city, surrounded by green hills, and the city is surrounded by water. “The city is in the landscape garden and the landscape garden is in the city.” We incorporate the concept of “dragon travels in the landscape” into the architectural shape, to make a contrast between the scattered building dimension and prolific facade, with forming a continuous line pattern on the facade of the building, abstracting the image of the dragon, creating an elegant and bright architectural shape, forming a unique city card on the Liujiang side.

东莞生态园控股有限公司办公楼工程设计

Office Building of Dongguan Eco-park Holding Co., Ltd. Project Design

项目位置：广东省东莞市
设计时间：2010 年
竣工时间：2012 年
建筑规模：37,664 m^2

Location: Dongguan, Guangdong
Design: 2010
Completion: 2012
GFA: 37,664 m^2

依据岭南建筑传统，本项目面水背山地布置于场地中央，大致为南北向。建筑采取分散的错列式布局，分为四个小塔楼，在首层以一裙楼整体相连，有利于在塔楼之间形成通风通道，改善通风效果。建筑的布局保证了建筑所有部分都有较好的朝向，结合相对较小的进深也保证了建筑在较长的时间段内能自然通风。建筑外侧设置穿孔铝板作为外遮阳构件，其所形成的韵律感形体成为该办公楼造型的主要标志之一。

According to the Lingnan traditional architecture, the project is arranged in the center of the site, facing the water and with mountains on the back, basically facing north and south. The building adopts a scattered staggered layout, divided into four small towers, connected by a podium on the first floor, which is conducive to the formation of ventilation channels between the towers and improve the ventilation effect. The layout of the building ensures that all parts of the building have a good orientation, and the relatively small depth ensures that the building can take advantage of natural ventilation for a long period of time. Perforated aluminum plate is set on the outside of the building as the external shading component. The rhythmic shape formed by the shading component has become one of the main symbols of office building modeling.

南宁九洲国际大厦
Nanning Jiuzhou International Building

项目位置：广西壮族自治区南宁市
设计时间：2010 年
竣工时间：2017 年
建筑规模：215,600.23 m^2

Location：Nanning, Guangxi
Design：2010
Completion：2017
GFA：215,600.23 m^2

项目基地位于广西南宁琅东新区东盟商贸核心 CBD，地理位置相当优越。项目总建筑面积 215,600.23 ㎡，容积率 8.5。采用单塔设计，地下 6 层，地上 71 层，建筑高度 318m，是集商业、酒店、办公等功能为一体的综合性超高层建筑。

项目用地狭窄，面积仅 1.15 万㎡，按规划要求需设 835 个停车位，为满足绿地率和停车要求，本项目设计了六层地下室，基坑深度达 25.1m。本项目采用钢管柱混凝土框筒结构，以 5560 元 / 平方米的综合建安造价，成为超过 300m 深基坑超高层的典范。

The project base is located in the CBD of the ASEAN Business and Trade Center in the New District of Nanning, Guangxi. The total construction area of the project is 215,600.23 square meters and the floor area ratio is 8.5. With a single tower design, 6 floors underground, 71 floors above ground, and a building height of 318 meters, it is a comprehensive super high-rise building integrating functions of business, hotel and office.

The project has a narrow land area of only 11,500 square meters. According to the planning requirements, 835 parking spaces are required. To meet the green space rate and parking requirements, the project has designed a six-story basement with a foundation pit depth of 25.1 meters. The project adopts the steel tube column concrete frame tube structure, with a comprehensive construction cost of 5560 yuan / square, which becomes a model of super high-rise of more than 300 meters deep foundation pit.

广州南站核心区 BA01077 地块设计（路福联合广场）

Block BA01077 (Rofo Union Plaza), Core Area of GuangZhou South Railway Station

项目位置：广东省广州市
设计时间：2013 年
竣工时间：2018 年
建筑规模：63,729 m^2

Location: Guangzhou, Guangdong
Design: 2013
Completion: 2018
GFA: 63,729 m^2

路福联合广场位于广州南站核心区，是路福房地产开发有限公司总部和国内首家丽芮酒店所在地，属典型的超高层商业综合体项目。

为创造更有趣生动的空间体验和更丰富的公共生活，同时实现经济性和公共性的平衡，项目利用避难层及屋顶空间，形成多处不同主题的空中公园，将户外空间和公共生活引入建筑内部，为使用者提供了一处具有富有特征和故事性的空间场景，也因此成就了“城市之眼”的建筑意象。

Located in the core area of Guangzhou South Railway Station which is the headquarters of Rofo Group and the seat of the first Radisson Red Hotel in China, Rofo Union Plaza is a typical super high-rise commercial complex project.

To create more interesting and vivid spatial experience and richer public life, and achieve a balance between economy and publicity simultaneously, the project takes advantage of refuge storey and roof spaces to form a variety of different themed aerial parks that introduce outdoor spaces and public life into the building interior. It provides the users with a story-filled space scene and creates the architectural image of the "eye of the city".

3990 7

kin coffee

武汉光电创新园（一期）
Wuhan Optoelectronic Innovation Park (Phase Ⅰ)

项目位置：湖北省武汉市
设计时间：2016 年
竣工时间：2019 年
建筑规模：90,713.06 m²

Location：Wuhan, Hubei
Design：2016
Completion：2019
GFA：90,713.06 m²

本设计以“智 · 汇光电，联 · 创光谷”为设计意向，以“智慧之芯，汇聚光电；联袂精英，创新光谷”为构思框架，通过“汇聚”“汇合”“荟萃”的抽象手法，把研发、中试、公共服务等空间整合成一个有机整体，为光电创新工作者提供一个智慧碰撞、思想迸发的众创乐园。

The design is based on the intention of “Intelligent Optoelectronic, Innovative Optics Valley” and the conceptual framework of “the core of intelligence gathering in optoelectronic, the elites innovating in Optics Valley”. Through the abstract methods of gathering, assembling and extracting, the spaces of R&D, pilot-scale experiment and public service have been integrated into an organic whole. We provide an innovative park for optoelectronic innovators to exchange and stimulate their thoughts.

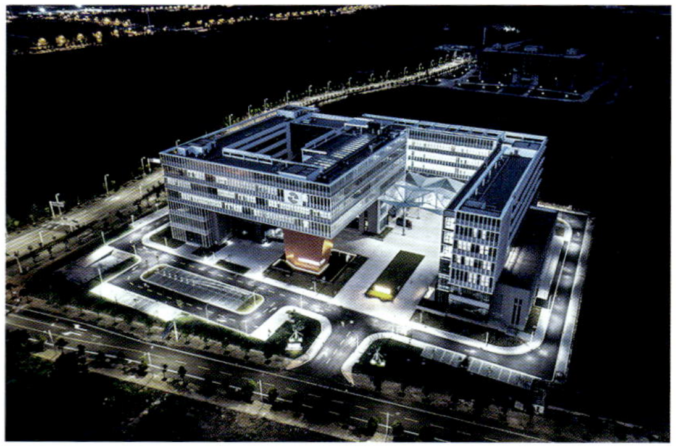

渔安·安井温泉旅游城“未来方舟”B5、B6组团

Yu'an Anjing Hot Spring Tourism Town “Future Ark” B5 and B6 Groups

项目位置：贵州省贵阳市
设计时间：2013—2016 年
竣工时间：2018 年
建筑规模：128,000 m^2

Location：Guiyang, Guizhou
Design：2013–2016
Completion：2018
GFA：128,000 m^2

渔安·安井温泉旅游城“未来方舟”B5、B6组团西接十六号规划道路，东临绕城高速，东西两侧平均高差约40m。项目用地东西延展50~75m，南北长约800m，基地沿南明河展开，用地狭长，地处山地且坡度较大，等高线成连续“Z”字形分布，有别于传统概念的坡地地貌。从生态建筑、生态环境的理念出发，项目单体顺应山地等高线走势布局，而后通过空间折板与“肌肉”状连桥将各单体建筑串联起来，不仅增加了建筑南北朝向面积，同时也扩大了建筑的景观面，形成个性独特、标识鲜明的“空中街道”。该项目具有发展商业、公寓的独家优势，在该地段发展集商业、公寓为一体的国际性商业街具有合理性。

Yu'an Anjing Hot Spring Tourism Town “Future Ark” B5 and B6 groups are connected to the planned No.16 Road in the west, and the east is beside the expressway. The average height difference between the east and west sides is about 40 meters. The project land extends 50-75 meters, and the north and south are about 800 meters long. The base is developed along the Nanming River. The land is long and narrow, and it is in the mountainous area with a large slope. The contour line is continuously “Z” shaped, which is different from the traditional landform. Starting from the concept of ecological architecture and ecological environment, the project unit is in line with the contours of the mountain contours, and then the individual buildings are connected in series through the space folding plate and the “muscle” bridge, which not only increase the north-south facing area of the building, but also expand the landscape of the building and form an “air street” with unique personality and distinction. The project has the exclusive advantage of developing commercial and apartment buildings, and it is reasonable to develop an international commercial street integrating business and apartments in this area.

3
4
6

2
3
4
5

东莞松山湖科技产业园区科学苑

Science Park, Gongguan Songshan Lake Science and Technology Industrial Park

项目位置：广东省东莞市
设计时间：2008 年
竣工时间：2010 年
建筑规模：120,400 m^2

Location: Dongguan, Guangdong
Design: 2008
Completion: 2010
GFA: 120,400 m^2

松山湖科技产业园科学苑是松山湖科技产业园区重要的标志性建筑组群之一。

科学苑的设计顺应松山湖整体规划思想，借鉴岭南园林的设计手法，创造宜人的岭南庭院尺度和空间模式；尊重现有地形地貌环境，保留山体，集中布局建筑，结合当地气候条件，采用节能技术手段；外部空间通过多层次园林化处理，形成指状绿化布局；打破各功能相互独立布局的模式，在独立的工作空间之外提供相互联系的共享平台和交往空间，促进学科交流、资源共享。最终形成一个兼具地域化、生态化、园林化、网络化的现代科技园区。

Science Park is one of the important landmark buildings of Songshan Lake Science and Technology Industrial Park.

In the design of Science Park, we follow the overall planning idea of Songshan Lake, draw lessons from the design method of Lingnan garden, and create the pleasant scale and spatial pattern of Lingnan garden. In addition, we respect the existing landform and environment, retain mountains, concentrate the layout of buildings, combine local climate conditions, adopt energy-saving technology. The outer space is treated with multi-level gardening and forms the finger-like greening layout. To break the pattern of independent layout of each function, we provide a shared platform and communication space for each outside the independent working space, and promote the exchange of disciplines and resource sharing. Finally, a modern science and technology park with regional, ecological, gardening and networking has been formed.

WIIO

WIIO

君豪国际商业城

Junhao International Business Town

项目位置：广东省广州市
设计时间：2013—2015 年
竣工时间：2016 年
建筑规模：169,000m^2

Location：Guangzhou, Guangdong
Design：2013-2015
Completion：2016
GFA：169,000 m^2

君豪国际商业城位于广州市著名的中大商业板块，在广州大道一侧的门户位置，独特的选址决定了其将成为整个板块的商业形象面。在实体商业饱受虚拟商业体冲击的今天，该项目在设计之初就承担着继承和展现这一板块的新的商业雄心的寄托。建筑在形式上追求速度感与流动性意象，通过数段笔挺的线条结合内部商业分区对裙楼商业立方体做出概念切分，结合内在的功能模块划分及环境需求，以设计外化为造型逻辑。运用理性分析及简约的笔触，在城市语境中表达尊重。为引入自然光线，君豪国际商业城的商业中庭采用了带形的天窗形式，使商业中庭遮阳系统的节能效能得以实现。

Junhao International Business Town is located at the gateway of Guangzhou' s famous SYSU commercial plate on the side of Guangzhou Avenue. The unique location determines that it will become the commercial image of the entire sector. Today, when physical business is hit by virtual commerce, the project has been designed to inherit and showcase the new business ambitions of this sector. The building pursues the sense of speed and fluidity in form. Through several straight lines combined with the internal commercial division to divide the concept of the podium commercial cube, and with the internal functional module division and environmental requirements, the modeling logic of the design is externalization. Rational analysis and simple brushstrokes have been used to express respect in urban context. In order to introduce natural light, the commercial atrium of Junhao International Business Town adopts a belt-shaped skylight form to realize the energy-saving effect of the shading system.

中国体育彩票广西营销展示中心

China Sports Lottery Guangxi Marketing Exhibition Center

项目位置：广西壮族自治区南宁市
设计时间：2012 年
竣工时间：2017 年
建筑规模：17,600 m²

Location: Nanning, Guangxi
Design: 2012
Completion: 2017
GFA: 17,600 m²

项目建设地址位于南宁市五象新区良庆大道以东，庆歌路以北，楞塘冲水系以南。项目总用地面积 19980 m²（约 30 亩），净用地面积 14652 m²（约 22 亩），总建筑面积 17600 m²。结构主体高度不超过 24m，为多层建筑。

中国体育彩票广西营销展示中心的方案设计从原有地形出发，并在广西传统干阑式建筑中提炼空间元素，通过架空、埋覆、悬挑等手法，塑造出如丘陵起伏般高低错落的建筑形态。

The project construction address is located in the east of Liangqing Avenue in Wuxiang New District, Nanning, north of Qingge Road and south of Qiangtang River. The total land area of the project is 19,980 square meters (about 30 acres), the net land area is 14,652 square meters (about 22 acres), and the total construction area is 17,600 square meters. The height of the structural body is below 24 meters and is a multi-story building.

The design of the China Sports Lottery Guangxi Marketing Exhibition Center starts from the original terrain and refines the space elements in the traditional dry-style buildings in Guangxi. Through the techniques of overhead, embedding and cantilevering, it creates a high and low building like the rolling hills.

仁恒滨海中心

Yanlord Marina Center

项目位置：广东省珠海市
合作单位：美国 NBBJ 建筑设计公司
设计时间：2008—2012 年
竣工时间：2016 年
建筑规模：307,000m²

Location: Zhuhai, Guangdong
Partner: NBBJ
Design: 2008-2012
Completion: 2016
GFA: 307,000 m²

仁恒滨海中心位于珠海市中心城区南部，沿情侣南路形成约 600 米带状项目。仁恒滨海中心位置优越，地处拱北黄金商业地带，紧邻拱北口岸，正对港珠澳大桥珠海与澳门的入口处，广珠轻轨拱北枢纽站步行即可到达地块，是珠海市最具商业价值的中心地段。仁恒滨海中心写字楼以帆船造型建筑，全铝制立面，兼具节能环保、保温隔热、阻挡紫外线等优势。

Yanlord Marina Center is located in the south of downtown Zhuhai, forming a belt of about 600 meters along Qinglv Road South. Yanlord Marina Center is well located in the most commercial center of Zhuhai and close to Gongbei Port, facing the entrance of Hong Kong-Zhuhai-Macao Bridge, as well as within walking distance to Gongbei Station of Guangzhou-Zhuhai Intercity Railway. Based on the prototype of sailboats, the office buildings are designed to be all-aluminum facade that combines advantages of energy saving, thermal insulation and UV blocking.

西樵纺织科技大厦
Xiqiao Textile Technology Building

项目位置：广东省佛山市
设计时间：2012 年
竣工时间：2018 年
建筑规模：32,685 m^2

Location：Foshan, Guangdong
Design：2012
Completion：2018
GFA：32,685 m^2

设计既考虑对西樵地区纺织产业集群中心地位形态的塑造，又注重经济效益。引入了纺织产业形象理念，平面设计与造型设计均结合办公建筑本体功能与纺织产业经营功能，塑造标志性形象与商业性使用相统一的建筑形态。通过建筑的竖向布置与立面设计，令塔楼尽可能高耸挺拔，完整统一，从体量上达到地标的作用。强烈的色彩半掩在灰色调的活动遮阳板下，理性却又隐含随机的美感。

The design not only considers the shaping of the central status of the textile industry cluster in Xiqiao area, but also pays attention to economic benefits. The image concept of the textile industry is introduced. The graphic design and the modeling design combine the functions of the office building and the textile industry, to create a unified architectural form of the iconic image and commercial use. Through the vertical arrangement and the facade design of the building, the tower is as tall and straight as possible, complete and unified, and reaches the landmark role from the perspective of dimension. The strong colors are half hidden under the gray visor of the sun visor, showing the rational but implied random beauty.

广发证券大厦

Guangfa Securities Headquarters

项目位置：广东省广州市
合作单位：耶格卡恩建筑设计有限公司
设计时间：2011 年
竣工时间：2018 年
建筑规模：158,263 m^2

Location：Guangzhou, Guangdong
Partner：Jaeger Kahlen Partner
Design：2011
Completion：2018
GFA：158,263 m^2

广发证券大厦位于广州珠江新城，是一座超高层建筑。项目区位的独特之处在于，它位于一条主干道的末端，与广阔的珠江公园相邻，因而作为城市的节点，人们可以从道路、公园以及各条城市轴线上看到这座 308m 高的塔楼。在 2018 年完工的全球最高建筑中，排名第 18 位；在广州已建成的摩天大楼中，排名第 9。

Guangfa Securities Headquarters is a super high building located in Zhujiang New Town district of Guangzhou, China. The location is peculiar in that it lies at the end of a terminating arterial road and near the Zhujiang Park. The 308-meter tall tower thus assumes a nodal location with high visibility from the road, the park and varied urban axes in Zhujiang New Town district. It ranked 18th of the tallest completed buildings of the world in 2018, and 9th of the tallest completed skyscrapers in Guangzhou.

深圳后海天佶湾国际广场项目

Shenzhen Houhai Tianji Bay International Plaza

项目位置：广东省深圳市
设计时间：2014 年
建筑规模：450,470 m²

Location：Shenzhen, Guangdong
Design：2014
GFA：450,470 m²

天佶湾国际广场位于深圳南山后海片区，距海岸线约 1.3km。整个项目用地呈矩形，东西宽 170m，南北约 150m，由内部两条十字形道路分割为四个地块。项目总共由 2 栋 268m 的办公酒店塔楼、1 栋 150m 的公寓塔楼和 24m 的商业裙楼组成。建筑综合体通过总平面布局寻求景观关系的最优解。在具体位置的选择上，2 栋较高的塔楼位于商业价值较高的地块，按对角线错开布置，使景观面和互相间距达到最大化。

Tianzhu Bay International Plaza is located in the Houhai area of Nanshan, Shenzhen, about 1.3 kilometers from the coastline. The entire project is rectangular in shape, 170 meters wide from east to west and 150 meters from north to south. It is divided into four plots by two internal cross-shaped roads. The project consists of two 268-meter office towers, a 150-meter apartment tower and a 24-metre commercial podium. The building complex seeks the optimal solution of the landscape relationship through the general layout. In the choice of specific location, the two tall towers are located on plots with higher commercial value, staggered diagonally to maximize the landscape and mutual spacing.

澳门大学横琴校区	New Campus of the University of Macao in Hengqin Island
华南理工大学广州国际校区	South China University of Technology (Guangzhou International Campus)
中山大学深圳校区	Sun Yat-Sen University (Shenzhen Campus)
上海大学校本部东区	East Campus of Shanghai University
中国资本市场学院	China Capital College
合肥工业大学宣城校区	Hefei University of Technology Xuancheng Campus
中山大学珠海校区	Sun Yat-Sen University (Zhuhai Campus)
南京工程学院江宁校区	Nanjing Institute of Technology Jiangning Campus
六盘水民族中学高中部	Liupanshui Minzu High School
河源市源城区特殊教育学校	Special Education School in Yuancheng, Heyuan City
常州田家炳高级中学	Changzhou Tianjiabing Experimental High School
广西东盟国际职业技术学院	Guangxi ASEAN International Vocational and Technical College
湖南工贸技师学院	Hunan Technician College of Industry and Commerce
武汉理工大学南湖校区图书馆	Wuhan University of Technology South Lake Library
钱学森图书馆	Qian Xuesen Library
北京工业大学图书馆	Library of Beijing University of Technology
李政道图书馆	Li Zhengdao Library
中山大学科技综合楼	Science and Technology Complex, Sun Yat-sen University
常州工学院图书馆	Library of Changzhou Institute of Technology
中山大学科技文化交流中心	Science and Culture Exchange Center of Sun Yat-Sen University

03 教育建筑
EDUCATIONAL ARCHITECTURE

南京工程学院文体综合馆	Nanjing Institute of Technology Comprehensive Gymnasium
武汉大学文科综合楼	Humanities Institute of Wuhan University
东莞职业技术学院图书馆	Dongguan Vocational and Technical College Library
福建工程学院新校区图书馆	Library of FuJian University of Technology
临沂大学艺、音、美综合体	Linyi University Art, Music and Beauty Complex
吉林大学前卫南区图书信息综合楼设计	Library and Information Complex, Jilin University Qianweinan Campus
武汉光电国家研究中心	Wuhan National Laboratory for Optoelectronics
肇庆学院艺术教学大楼	Zhaoqing College Art Teaching Building
中国石油大学（北京）实验办公综合楼	China University of Petroleum (Beijing) Experimental Office Complex
常州市天合国际学校外国语学校综合体	Changzhou Tianhe International School Foreign Language School Complex
湖南女子学院逸夫图书馆	Shaw Library of Hunan Women's University
中国地质大学（北京）校园总体规划方案设计	China University of Geosciences (Beijing) Campus Conceptual Master Plan
广东药科大学云浮校区	Guangdong Pharmaceutical University Yunfu Campus
天津科技大学泰达校区规划设计	Planning and Design for TEDA Campus of Tianjin University of Science & Technology (TUST)
莆田学院新校区修建性详细规划	Detailed Planning of the New Campus of Putian College
方太大学	Fotile University
海南大学海甸校区更新改造	Hainan University Haidian Campus
前海合作区前湾十单元学校项目	School in Unit 10 of Qianwan, Qianhai New District
广东以色列理工学院	Guangdong Israel Institute of Technology
柳州市委党校搬迁工程	Party School of CPC Liuzhou City Committe
南京航空航天大学天目湖校区	Tianmuhu Campus of Nanjing University of Aeronautics and Astronautics

澳门大学横琴校区

New Campus of the University of Macao in Hengqin Island

项目位置：广东省珠海市
设计时间：2009 年
竣工时间：2012 年
建筑规模：816,000 m^2

Location：Zhuhai, Guangdong
Design：2009
Completion：2012
GFA：816,000 m^2

澳门大学横琴校区立足于用地的山海环境资源和自然条件，融合南欧和华南地区的建筑风格，凸显澳门作为中西文化交流平台之角色，打造富有独特魅力和风格的国际化校园和多层次、多中心的立体化、书院式发展的生态、人文校园。

围绕校园中心区形成多个组团相互围合的关系，形成大功能组团围绕生态湖泊和园林的公共开敞空间总体形态。各建筑组团形成书院式的发展格局，各书院形成各自的立体化中心与多层次步行系统，形成独立高效的运作系统。以舒适的步行尺度 300 m 为依据控制组团的大小，通过步行轴线联系各组团，形成完善的步行系统。

Based on mountain and ocean resources as well as the superb natural conditions, the University of Macao integrates architectural styles of South Europe and South China, highlighting the Macao's role as a platform of Chinese and western cultural exchange and creating an international campus with unique charm and style.

As planned, building clusters are placed around the campus center to mutually enclose with each other, thus form a general layout of the public open spaces where the large functional clusters embrace the eco-lake and gardens. Each building cluster is organized into a pattern that resembles the traditional Chinese academy architecture, while each academy forms a solid center and multi-leveled walking system of its own and independent and effective operation system as well. The building clusters are sized based on 300 meters, which is the comfortable walking distance. A complete walking system is created to connect the clusters by walking axes.

华南理工大学广州国际校区

South China University of Technology (Guangzhou International Campus)

项目位置：广东省广州市
设计时间：2017 年
竣工时间：2019 年
建筑规模：1,400,000 m^2

Location：Guangzhou, Guangdong
Design：2017
Completion：2019
GFA：1,400,000 m^2

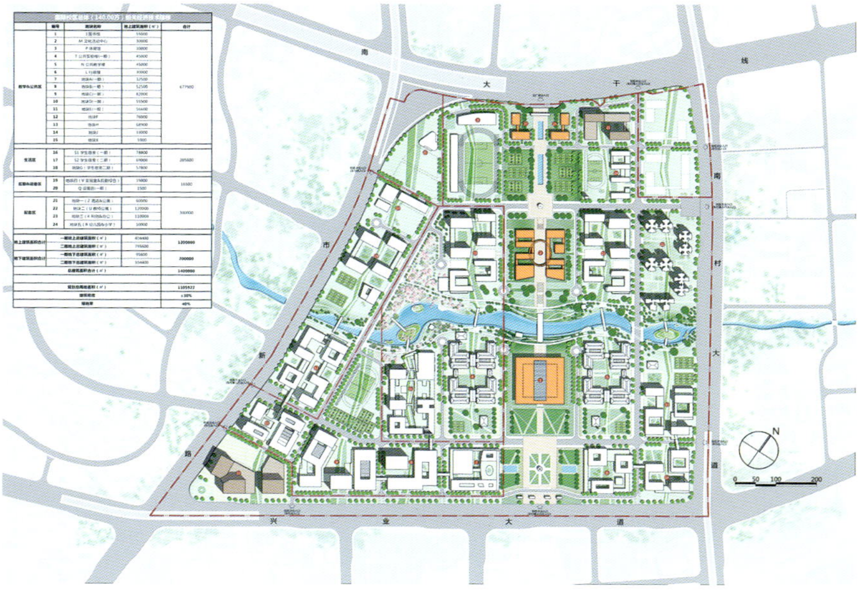

华南理工大学广州国际校区以“合璧”作为校园总体规划理念，紧扣“中西合璧、博雅合璧、传统与现代合璧、科学与人文合璧”的主旨，一方面提取和延续华南理工大学老校区的环境基因，另一方面再现与注入国外著名大学的空间意向元素，构建具有岭南地域特点并符合时代精神的综合性大学国际校区。

Taking “combination” as the main idea of campus master planning, South China University of Technology (Guangzhou International Campus) closely adheres to the purport of “combination of Chinese and western, combination of broadness and refinement, combination of tradition and modern, combination of science and humanities”. On one hand, it extracts and continues the environmental genes of the old campus of South China University of Technology. On the other hand, it reproduces and injects the spatial intention elements of foreign famous universities, so as to construct the international campus of a comprehensive university with Lingnan regional characteristics and in line with the zeitgeist.

中山大学深圳校区
Sun Yat-Sen University (Shenzhen Campus)

项目位置：广东省深圳市
设计时间：2017 年
竣工时间：2022 年
建筑规模：200,000,000 m^2

Location：Shenzhen, Guangdong
Design：2017
Completion：2022
GFA：2,000,000 m^2

上海大学校本部东区

East Campus of Shanghai University

项目位置：上海市
设计时间：2005 —2012 年
竣工时间：2013 年
建筑规模：270,000 m^2

Location：Shanghai
Design：2005-2012
Completion：2013
GFA：270,000 m^2

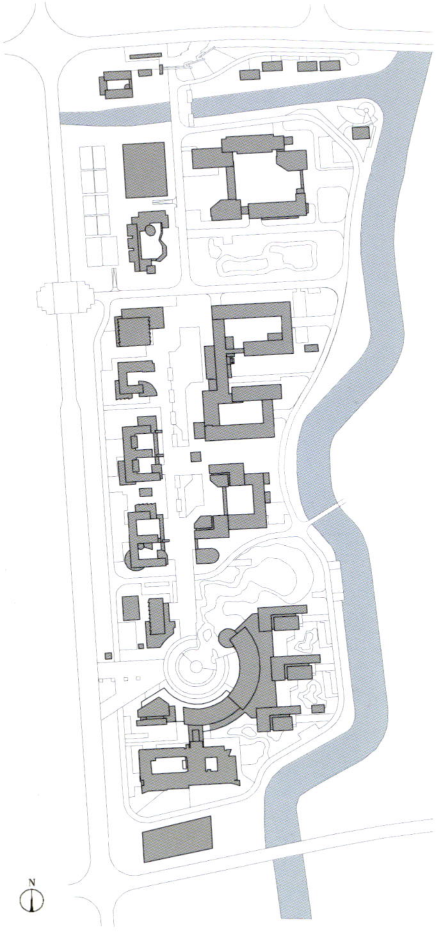

中山大学深圳校区位于光明区，校园一期用地面积 150 hm^2，以医科和工科为主，可容纳两万学生。用地一山矗立，七丘拱卫，新陂头河横贯其中，我们希望这是一所与山水相融的、有利于学术创新的大学校园。

中心山体作为校园的生态绿核，外部生态环境通过指状绿廊引入，建筑嵌在绿带之间。大尺度、高层的建筑组团布置在外围，点式、低层的建筑沿山布置，实现建筑与生态相融。

北面沿岸景观优美，规划把体育馆、活动中心和景观中心布置在这里，形成滨河活力带。同时结合两条雨洪廊道，形成海绵校园的主要骨架。

校园中心由南至北，沿轴线布置了牌坊、中山像、图书馆、礼堂和惺亭，既传承了中大历史文脉，又建立了校园空间秩序。

东西向两条步行带将学科组团、公共教学设施、宿舍串联起来。学科组团和公共教学楼相邻布置，强调学科交叉、教研融合。

Sun Yat-sen University (Shenzhen Campus) in Guangming District is planned with a land area of 150 hectares for Phase I development. With medicine and engineering as key disciplines, the campus is designed to accommodate 20,000 students. The site has a central mountain surrounded by seven smaller hills, with the Xinpitou River winding through. Our goal is to create a university campus that well blends with the landscape and inspires academic innovation.

The central mountain serves as the green core of the campus, while the external green corridors are brought into the site by the finger-like green corridors. The buildings are embedded in between the green belts. Larger and higher buildings are placed at the periphery while the lower ones and towers along the mountains, ensuring perfect integration with the eco-environment.

The waterfronts in northern part of the site enjoys attractive natural scenery, where the gymnasium, activity center and landscape center are planned to create a dynamic waterfront area. This waterfront area combines two stormwater corridors, forming the backbones of a sponge campus.

At the center of the campus, a Paifang (Chinese archway), a statue of Dr. Sun Yat-Sen, a library, an auditorium and the Xing Pavilion are placed along the central axis from south to north. This is a tribute paid to the tradition of the university and an approach to establish the spatial order of the campus.

Two east-west pedestrian belts connect discipline building clusters, public teaching buildings and dormitories. The discipline building clusters and public teaching buildings are placed side by side to facilitate the crossover of disciplines and integration of teaching and research.

中山大学图书馆

项目将园林化设计理念与创新型研究教学模式相结合，以多层次交往园林为中心，设计园林式学院群，以及有利于学科交叉和创新交往的建筑空间和适应研究生培养模式的功能布局，营造园林化的研究生教学区。

设计提出中心园林—共享平台—研究社区的模式：总体布局采用有利于学科交叉、资源共享的学科组团式布局，改善各专业封闭独立的传统布局，将相近学科的学院以共享园林为中心相邻布置，并在共享空间中设计交往园林和公共资料中心、学科图书馆等公共平台，实现相近学科的资源共享。

The campus planning combines the garden concepts and the innovative R&D teaching mode, and centers on the multi-level communication gardens to create the garden-like clusters of campus. Meanwhile, efforts are made to create the building spaces that encourage the cross-disciplinary exchanges and innovative communications, the functional layout that meets the needs of post-graduate education, as well as the garden-like post-graduate teaching area.

The campus planning proposes the mode of central garden-shared platform-research community. The master plan adopts a clustered layout to facilitate the cross-disciplinary exchanges and sharing of resources, and improves the traditional layout which isolates different faculties from each other. With a shared garden at the center, faculties focusing on similar disciplines are located next to each other. Public platforms such as gardens for communications, public information center and libraries for disciplines are also included in the shared spaces to realize information sharing between similar disciplines.

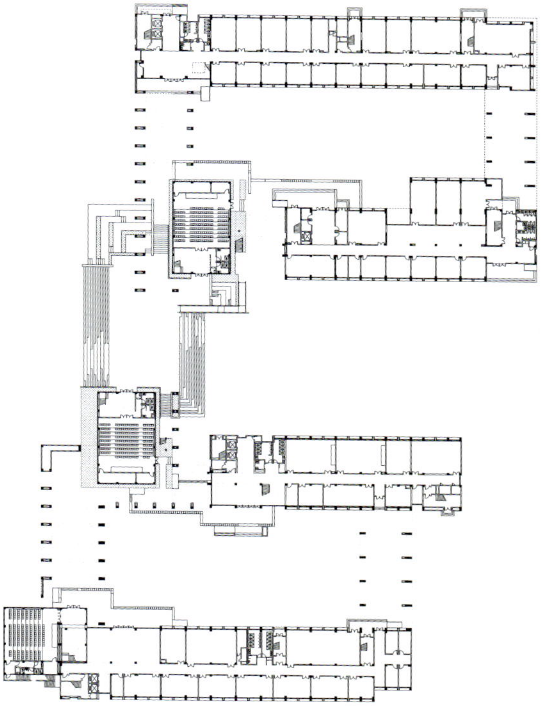

中国资本市场学院
China Capital College

项目位置：广东省深圳市
设计时间：2014 年
竣工时间：2017 年
建筑规模：20,129 m^2

Location：Shenzhen, Guangdong
Design：2014
Completion：2017
GFA：20,129 m^2

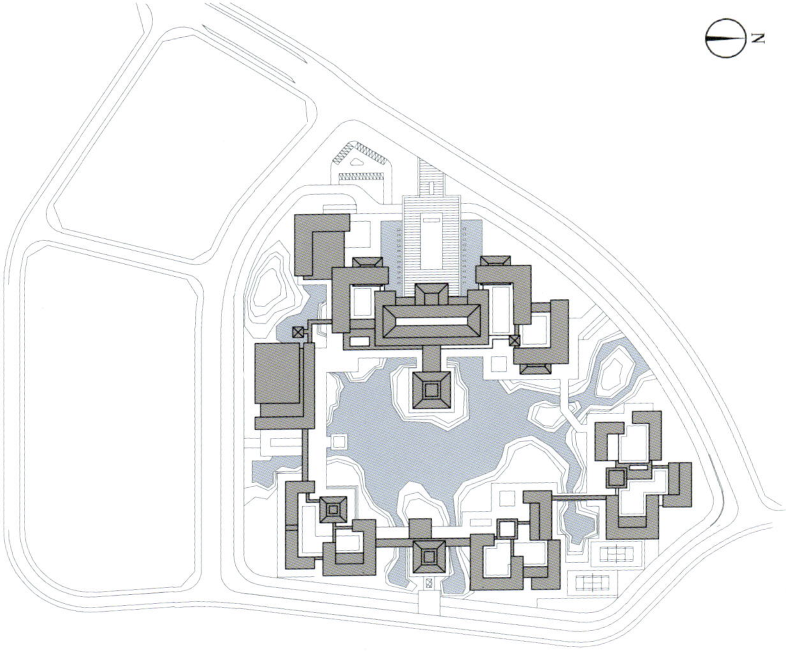

场地北靠麒麟山，场地中建筑成组团布置，组团间山水景互相渗透。西侧从水库引水入园，汇聚于校园中心，以此成为校园融山聚水的核心景观格局，寓意资本学院将吸引汇聚大批来自全国各地的高端人才，并推动整个资本市场的未来发展。

设计注重创新型多样化的交往空间的营造。创新思维的产生与创新知识的传播有赖于交往行为。交往空间特别是非正式交往空间，是促进创新产生的重要条件。山、水、建筑与园林空间互相结合，形成多层次人性化的园林式布局，能够为教学之余的自由交流提供各种舒适的场所。

建筑形象力图塑造具有中国传统文化神韵而又典雅庄重的现代校园建筑群。平坡结合的建筑群体造型体现了时代精神与传统气质。石材、钢、玻璃幕墙等现代的建筑材料与传统园林式布局相结合，表现出整体典雅大气的校园群体形象。同时，将数字化、绿色生态、可持续发展的理念融入建筑的规划、设计和未来的运营管理中。我们以网络化、可生长式的建筑布局实现校园的弹性发展，同时综合运用各种低碳节能环保策略、数字化管理策略打造绿色、节能、智慧校园，营造出一种人与自然和谐共生的学习生活环境。

The site sits against Qilin Mountain on the north, while the buildings on the site are planned into clustered layout with mountain and water features in between. Water is diverted into the center of the campus from the reservoir on the west of the site, realizing the central concept of landscaping, bringing mountain and water elements into the campus. It is a metaphor for the Capital College's mission of attracting top talents from all over the country and promoting the future development of the entire capital market.

We intend to create innovative and diversified communication spaces in the design, as the generation of innovative thinking and the spread of innovative knowledge depend on communicative behaviors. Communication spaces, especially informal spaces, is an important factor to promote innovation. Here mountains, water elements, buildings and garden spaces interact with each other, together making up a multi-level and human friendly garden layout that can provide various comfortable spaces for free communication outside classroom.

In terms of building image, the goal is to create a modern campus building complex with both the charm of Chinese traditional culture and solemn elegance. The building clusters adopt both flat roofs and pitched ones, displaying both the spirit of the times and traditional temperament. Modern building materials, such as stone, steel and glass curtain wall, and the traditional garden layout are brought together to create an elegant and graceful image of the campus. New concepts, such as digitalization, green ecology and sustainable development are also integrate into the plan, design and future operation of the buildings. We use the digitalized and sustainable buildings to achieve the resilient campus development, and adopt various low-carbon and energy-saving environmental protection strategies and digitalized management strategies to create a green, energy-saving and smart campus, ensuring a living and learning environment where human and nature co-exist harmoniously .

合肥工业大学宣城校区

Hefei University of Technology Xuancheng Campus

项目位置：安徽省宣城市
设计时间：2012 年
竣工时间：2016 年
建筑规模：609,400 m^2

Location：Xuancheng, Anhui
Design：2012
Completion：2016
GFA：609,400 m^2

方案通过依山就势的生态布局来表达一个现代高校与水转山绕的自然格局、人文荟萃的地方传统相结合的诗意画面，在山体水系的基础上，形成了“三山四水，岛翠墨香”的独特格局，山成为校园的背景和绿肺，水系及其滨水区域则为整个校园景观及活动中心。

建筑群体以顺应、环绕的姿态，更强化了基地优美的地表特征。而岛状组团之间的冲沟又将背景山麓的自然景色强有力地渗透到校园中心，保持了生态景观的连续性。

方案通过理性的分析与推敲，又叠加了有机的空间架构关系，将有序的建筑融入有机的大自然。方案建立了东西和南北两条轴线，控制了整个校园的序列格局；各个区块通过灵活的体量组合和轴线变化，形成了灵活多样的空间体验。

Through the ecological layout that conforming to the mountain, the program expresses a poetic image of a modern university with a natural pattern of water and mountains, and a local tradition of humanities. On the basis of the mountain and water system, it forms the unique pattern of “three mountains and four rivers, green island with the fragrance of ink” . The mountain becomes the background of the campus and the green lungs, and the water system and its waterfront area are the entire campus landscape and activity center.

The building group adopts a conforming and surrounding posture, which further strengthens the beautiful surface features of the base. The gully between the island groups strongly infiltrates the natural scenery of the background mountain into the center of the campus, maintaining the continuity of the ecological landscape.

Through rational analysis and refinement, the scheme superimposes the organic spatial structure relationship and integrates the orderly architecture into the organic nature. The program establishes two axes of east-west and north-south, which controls the sequence pattern of the whole campus; each block forms a flexible and diverse space experience through flexible dimension combination and axis change.

中山大学珠海校区

Sun Yat-Sen University (Zhuhai Campus)

项目位置：广东省珠海市
设计时间：2016 年
竣工时间：2021 年
建筑规模：1,600,000 m^2

Location：Zhuhai, Guangdong
Design：2016
Completion：2021
GFA：1,600,000 m^2

六盘水民族中学高中部

Liupanshui National High School

项目位置：贵州省六盘水市
设计时间：2014 年
竣工时间：2016 年
建筑规模：159,679 m^2

Location：Liupanshui, Guizhou
Design：2014
Completion：2016
GFA：159,679 m^2

六盘水民族中学高中部位于教育城东南侧，为一所能容纳6000 人、120 班的省示范高中。学校片区为山地、丘陵地区，地势起伏明显。为了减少对自然的破坏和节省建设开支，本设计并未大面积地改造地形，而是在遵循尊重自然、尊重本土特色的前提下进行，打造与自然融为一体、具有地域特色的人文建筑景观。在建筑布局上，整个设计并没有刻意追求轴线，布局以顺势为法，空间以错落为趣。

Liupanshui National High School is a provincial demonstration high school of 6,000 students and 120 classes located in the southeast of Education City. The site is mountainous and hilly, and the terrain is obviously undulating. In order to reduce the damage to nature and to save construction costs, we do not transform the terrain on a large scale, but create humanistic architecture and landscape with regional characteristics, on the premise of respecting nature and local characteristics.In the layout, the entire design do not deliberately obey the axis. The layout is based on the trend, and the space is staggered and interesting.

河源市源城区特殊教育学校

Special Education School in Yuancheng, Heyuan City

项目位置：广东省河源市
设计时间：2017 年
竣工时间：2018 年
建筑规模：8,792m^2

Location：Heyuan, Guangdong
Design：2017
Completion：2018
GFA：8,792 m^2

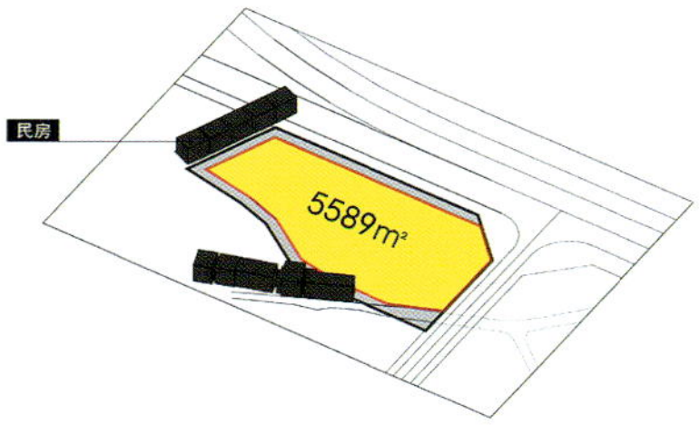

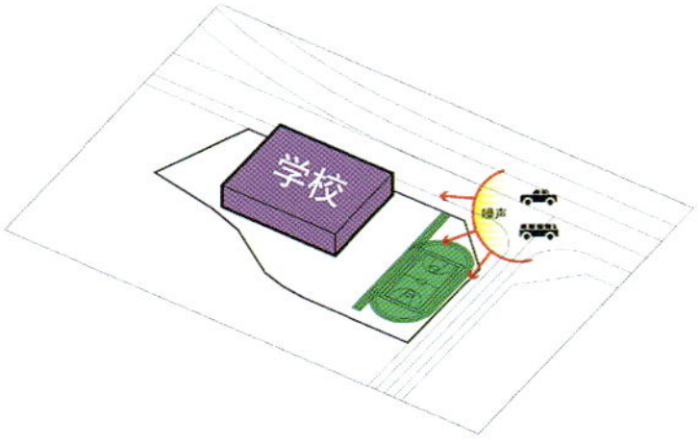

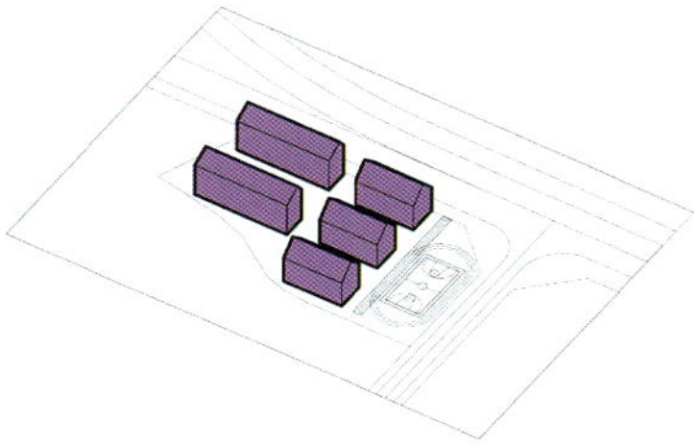

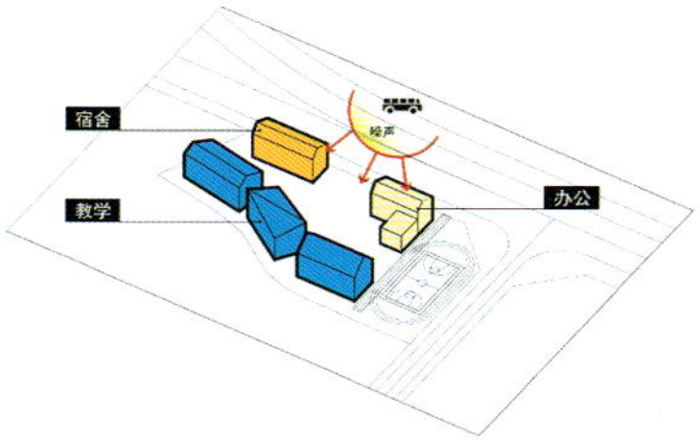

1. 村中“村”

项目设计学位 126 个，学生年龄段为 6~12 岁。基地处于一片低层高密度的民居之中，招收的孩子也大多来自于此。自由紧密排布的两三层小房屋、曲折的村道是孩子们最熟悉的元素。为此，设计从“村落”的原型入手，将学校按照使用功能拆分为若干个小体量建筑。各体量顺应用地边线自由布置，尽可能大地围合成内向的活动院落。低层高密度的校园空间特质、灵活的体块布局与周边村落相得益彰。从村落到学校，空间尺度自然过渡，增强学生对校园空间的认同感。

2. 校中“家”

随着城市的发展，周边民居命运未知，学校的建筑语言力图为孩子们保留最初“家”的记忆。各功能体量采用不同的坡顶形式，形成独一无二的小房子，增强标识性；在立面设计方面，建筑外墙以白色为主，与周边浅色调的民居呼应，东西山墙面也采取与民居相同尺度的方洞，自由散落在白墙之上，在体现民居特色之余不乏现代感。各小房子通过自然柔和的连廊联系为一体，好似蜿蜒的村道般在建筑之间流动。

i. The “village” in the village

The project has 126 designed degrees, in which its students are aged six to twelve. The base is in a high-density residential area at the bottom, and most of the children recruited by the school are from it. Two or three stories of closely related small houses and the winding village road are the elements most familiar to children. Therefore, the design starts with the prototype of “village”, so the school is divided into a number of small buildings according to its function. According to the free arrangement of the boundary line of the land, the inwardly activity courtyard is made as much as possible. The low level and high density campus space characteristic, the flexible layout and the surrounding village complete each other. From village to school, the natural transition of spatial scale can strengthen students' identity to campus space.

ii. The “home” in the school

With the development of the city, the fate of the surrounding residents is unknown. The architectural language of the school tries to preserve the original memory of “home” for the children. For each function dimension, it uses different slope top forms, which forms the unique small house, so as to enhance the marking. In the aspect of facade design, the exterior wall of the building is mainly white and completes the shallow hues of the surrounding residential buildings. The east and west walls also adopt square holes of the same scale as the residential buildings, which are scattered freely on the white walls. There is full of modern feeling while reflecting the characteristics of the houses. The small houses, connected by natural and gentle corridors, flow between buildings like winding village roads.

南京工程学院江宁校区
Nanjing Institute of Technology Jiangning Campus

项目位置：江苏省南京市
设计时间：2002—2009 年
竣工时间：2012 年
建筑规模：1,330,000 m^2

Location：Nanjing, Jiangsu
Design：2002–2009
Completion：2012
GFA：1,330,000 m^2

南京工程学院规划设计基于现有地形地貌特点，采用了自由园林式的生态校园布局方式，尊重自然、保护自然并创造性地利用自然。在校园发展模式上，主张簇群式发展的弹性生长校园模式，使新校区规划成为一个不断生长的动态发展过程。这种模式适应学校分期发展的特点，保证了教育设施的多用途并提高其使用效率。在单体设计风格上，主张体现地域性的现代化工科建筑风格，单体设计手法方面强调了简洁的体量与细腻的钢结构之间的对比，整体的效果和谐而富于变化。

Based on the existing topographic and geomorphological characteristics, the planning and design of Nanjing Institute of Technology Jiangning Campus adopts a free garden-style ecological campus layout, respecting nature, protecting nature, and using nature creatively. In the development mode of campus, the flexible growth mode of cluster development is advocated, to make the new campus planning into a growing dynamic development process. This model adapts to the characteristics of school development by stages, ensures the multi-purpose of educational facilities and improves their efficiency. In terms of monomer design style, it advocates the regional modern engineering architectural style. In terms of monomer design technique, it emphasizes the contrast between simple dimension and delicate steel structure, and the overall effect is harmonious and full of changes.

项目围绕建设世界一流大学的战略目标，规划创新复合的学科布局，营造理性浪漫的活力场所，构建文脉传承的经典空间，打造倚山临海的校园格局。

各学科群内布置有利于促进学科交叉的共享平台，公共建筑如图书馆、博物馆、会堂等散布其间，形成校园活力空间与创新网络。

在采用先进绿色技术的同时，引入中大百年老校的经典场所特质，以长轴串联中山像、牌坊、草坪等元素形成多进式空间。

Centering on the strategic goal of building a world-class university, the design aims to plan an innovative and comprehensive layout for discipline development, a dynamic place with rational romance, a classic space allowing for inheritance of culture and tradition, and a campus embracing mountains and sea.

Shared platforms that promote cross-disciplinary interactions are provided in every discipline building cluster, while public buildings, such as libraries, museums and auditoriums are scattered in between, establishing a dynamic spatial and innovative network on the campus.

While employing up-to-date green technologies, elements of classic landmarks of the century-old university are referenced as well. The statue of Dr. Sun Yat-Sen, a Paifang (Chinese archway) and lawns are linked up in sequence with a long axis, forming a multi-sequence space.

常州田家炳高级中学
Changzhou Tianjiabing Experimental High School

项目位置：江苏省常州市
设计时间：2014 年
竣工时间：2017 年
建筑规模：70,962 m^2

Location：Changzhou, Jiangsu
Design：2014
Completion：2017
GFA：70,962 m^2

常州市田家炳高级中学位于古运河畔、同济桥塊，设计灵感来源于中国传统建筑群院落空间，以方格网和轴线来控制建筑群体的生成，形成有序的建筑秩序。本案借鉴中国传统规划布局方式，采用“一心两轴多点”的形式来组织整个校区的规划设计，通过轴线、对称、均衡等手法强化校园的秩序感，使整个校区规划布局疏密有致、虚实相生、秩序井然。

Changzhou Tianjiabing High School is located on the bank of the ancient canal and Tongji Bridge. The design is inspired by the courtyard space of traditional Chinese buildings. The square grid and the axis control the formation of the building group and form an orderly architectural order. This case draws on the traditional Chinese planning and layout method, adopts the form of “one core, two axes and multiple points” to organize the planning and design of the entire campus, and strengthens the sense of order of the campus through the methods of axis, symmetry and balance, so that the layout of the entire campus is coordinated and orderly.

广西东盟国际职业技术学院

Guangxi ASEAN International Vocational and Technical College

项目位置：广西壮族自治区崇左市
设计时间：2017 年
竣工时间：2019 年
建筑规模：257,537.21 m^2

Location：Chongzuo, Guangxi
Design：2017
Completion：2019
GFA：257,537.21 m^2

设计以缔造东盟特色，借鉴中国传统合院式空间布局和重檐与架空等东盟建筑特征，以轴线、对称、均衡等手法强化校园的空间序列，以规划、景观、建筑三位一体的整体化校园设计手法，打造壮域学府、东盟之窗。

“一轴”是指校园中心礼仪轴。规划中充分考虑主入口的形象问题，在东面设置校园主入口，并于主入口处设校前广场，两侧的教学楼围合成校园的礼仪广场，并延伸至图书馆，形成一条中心礼仪主轴。“一带”是指沿湖景观带。用地中心设置连续的景观水体，各大功能区环湖而设，有效地串联起园区功能组团，形成一条贯穿南北的沿湖景观带。

The design aims to create the characteristics of ASEAN, draws on the traditional Chinese courtyard-style spatial layout and the characteristics of ASEAN architecture such as heavy and overhead, strengthens the spatial sequence of the campus by means of axis, symmetry and balance, and integrates the campus design with the trinity of planning, landscape and architecture. To build a strong university and the window of ASEAN is a main goal.

“One Axis” refers to the campus center etiquette axis. In the planning, the image of the main entrance is fully considered. The main entrance of the campus is set up in the east, and the pre-school square is set up at the main entrance. The teaching buildings on both sides are integrated into the campus etiquette square and extended to the library to form a central etiquette spindle. “A Zone” refers to the landscape along the lake. The continuous landscape lake is set in the center of the land, and the major functional areas are set up around the lake. The functional groups of the park are effectively connected in series to form a landscape belt along the north and south.

湖南工贸技师学院

Hunan Technician College of Industry and Commerce

项目位置：湖南省株洲市
设计时间：2007 年
竣工时间：2012 年
建筑规模：150,000 m^2

Location: Zhuzhou, Hunan
Design: 2007
Completion: 2012
GFA: 150,000 m^2

伴随着城市扩张以及高校教育的普及，很多高校因用地规模不足而将新校区选址在城市新区，高校集聚在城市新区形成大学城。湖南工贸技师学院位于株洲职业教育大学城西北，距离老城约 10 km，北临长沙，西接湘潭。学院规划总用地 16 hm^2，建筑面积约 15 万m^2，规划在校生为 8000 人。

一体化的校园设计强调校园布局、交通动线、功能空间和室外景观等要素的综合交叉和联结渗透。建筑与城市环境是一个有机的整体，只有有序的空间形态才能与城市设计形成良好的互动关系。加强室内和室外的空间联系，将建筑与景观的边界模糊，形成一种整体和谐的人性化空间。因此一体化设计更有助于创造紧凑高效的校园环境，同时促进交叉学科的发展和人才的培养。

With the expansion of cities and the popularization of higher education, many colleges and universities have chosen new campus sites in new urban districts because of the lack of land use. Colleges and universities gather in new urban districts to form university towns. Hunan Technician College of Industry and Commerce is located in the northwest of Zhuzhou Vocational Education University City, about 10 kilometers from the old city, north of Changsha, west of Xiangtan. The college has a total land area of 16 hectares and a construction area of about 150,000 hectares, with 8,000 students planned.

The integrated campus design emphasizes the comprehensive intersection and infiltration of elements such as campus layout, traffic lines, functional space and outdoor landscape. Architecture and urban environment are an organic whole. Only an orderly spatial form can form a good interaction with urban design. The design strengthens the space connection between indoor and outdoor, blurs the boundary between architecture and landscape, and forms a kind of holistic and harmonious humanized space. Therefore, integrated design is more conducive to creating a compact and efficient campus environment, while promoting the development of cross-disciplinary and personnel training.

武汉理工大学南湖校区图书馆

Wuhan University of Technology South Lake Library

项目位置：湖北省武汉市
设计时间：2010—2015 年
竣工时间：2016 年
建筑规模：47,557.6 m²

Location：Wuhan, Hubei
Design：2010–2015
Completion：2016
GFA：47,557.6 m²

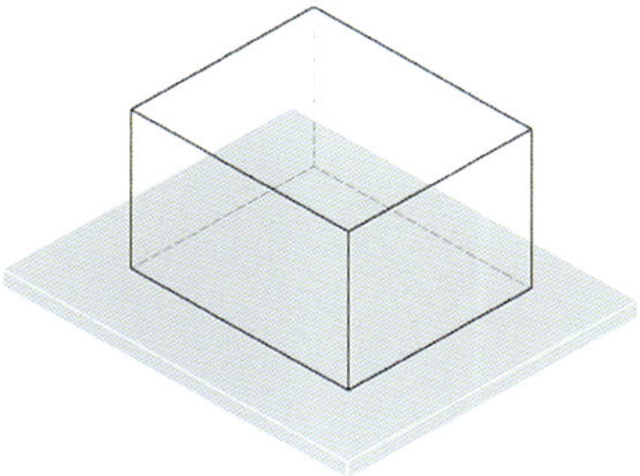

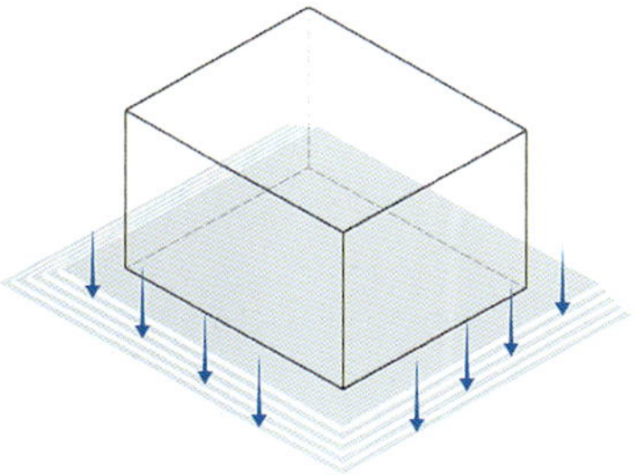

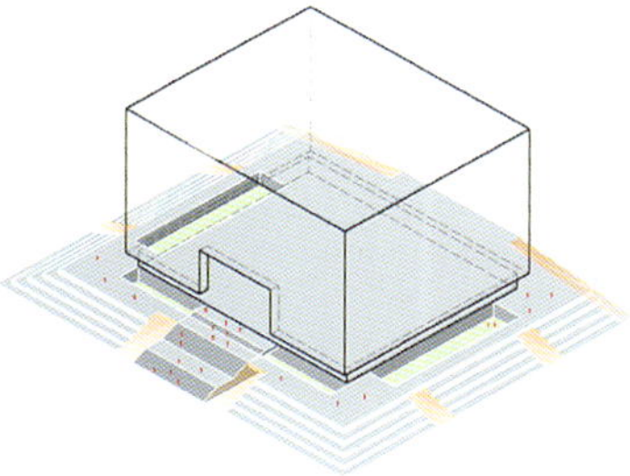

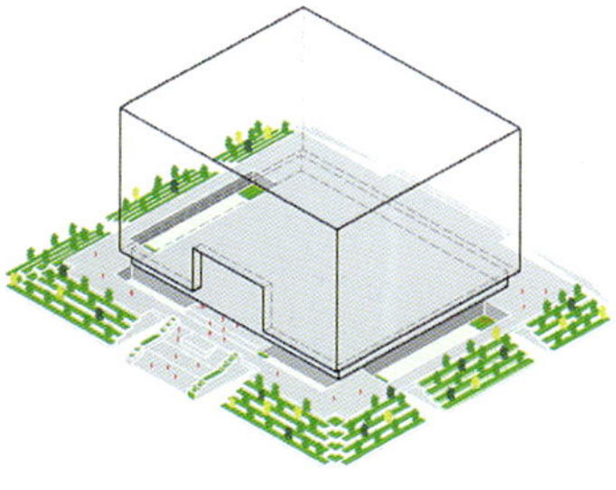

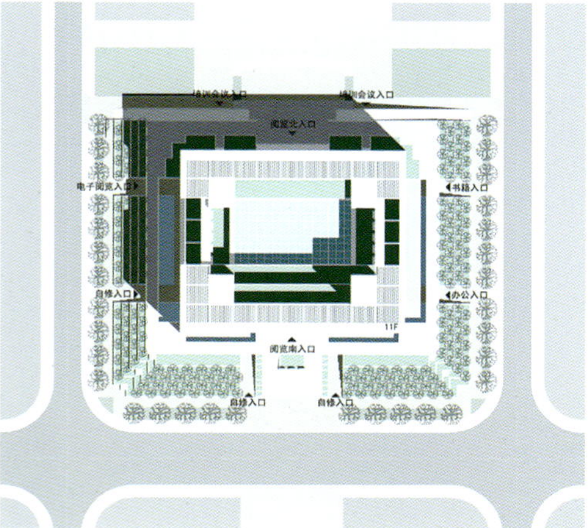

钱学森图书馆
Qian Xuesen Library

项目位置：上海市
设计时间：2009 —2010 年
竣工时间：2011 年
建筑规模：7,960 m^2

Location：Shanghai
Design：2009–2010
Completion：2011
GFA：7,960 m^2

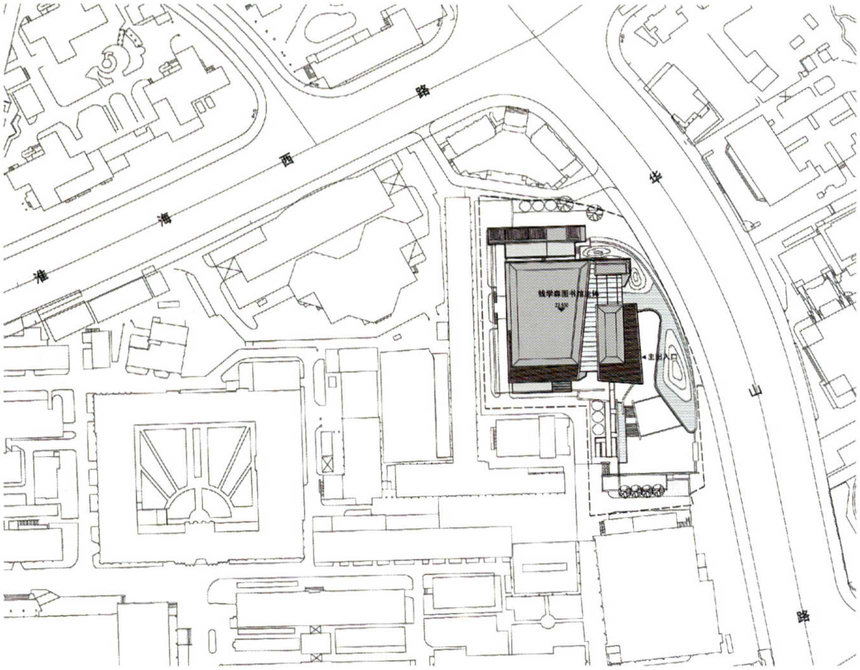

1. 楚风汉韵

（1）筑楚台

图书馆位于平坦宽阔的轴线中心，是整个校园的标志性建筑。建筑必须有足够大的体量创造具有控制力的广场界面。因此，设计将首层设计为基座——五级的跌台绿化，图书馆主体立于高台之上。

（2）传统木构件的现代演绎

图书馆主入口通过对传统斗拱、排架的拆分和重组，以钢结构构架演绎传统木构的形态。东西立面覆以木色金属板，并将校园历史文化通过抽象篆书文字镂刻其上。将楚文化进行抽象化和演绎，使历史底蕴在新建筑上得以延续和传承。

2. 书山绿谷

设计提取传统院落的精粹，以立体庭院的方式营造绿谷的意向。玻璃中庭悬挂绿色垂幔，形成对直射阳光的有效遮挡，也营造出绿意盎然的中庭效果。同时在首层基座跌台绿化设置雾喷，改善图书馆建筑的微气候。

Ⅰ. Ethos of Chu & Charm of Han

ⅰ. Construct high-platform building as Chu's charm

The library located in the center of the flat and broad axis, which as the landmark of the whole campus. The building must be large enough to create a controlled square interface. Therefore, the first floor is designed as a foundation base which have five-level greening, the main body of the library on top of the platform.

ⅱ. Modern expression of traditional wooden frame

The main entrance of library uses the steel structure frame to interpret the traditional wood structure by disassembly and reorganization of the traditional Dougong and brackets. The east and the west elevation covered with wood colored metal plates and carved with seal character which abstract from campus history and culture. Abstract and deduce the Chu culture, continue and inherit the historical background in the new building.

Ⅱ. Book mountain & Green valley

The design essences the extraction of traditional courtyard, and displays a green valley intention by stereoscopic courtyard. The glass atrium hangs green mantle to shield direct sunlight effectively, also creates atrium atmosphere full of green and vigour. At the same time, the fogging fountain in the first floor foundation base afforestation help to improve the microclimate of library buildings.

交通大学
钱学森图书馆

战略科学家的成功之道
EY TO SUCCESS OF A STRATEGIC SCIENTIST

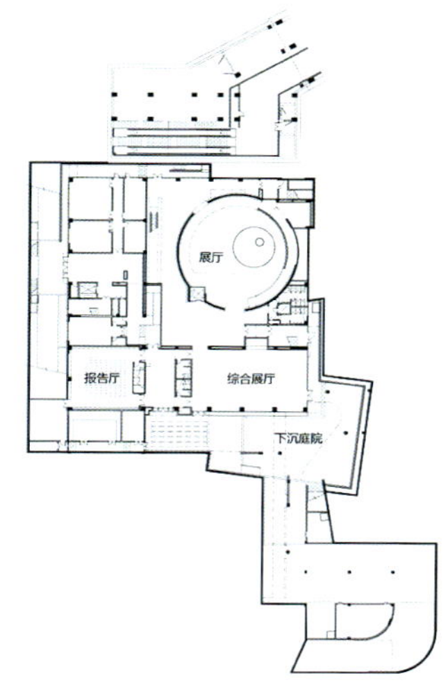

展厅
报告厅
综合展厅
下沉庭院

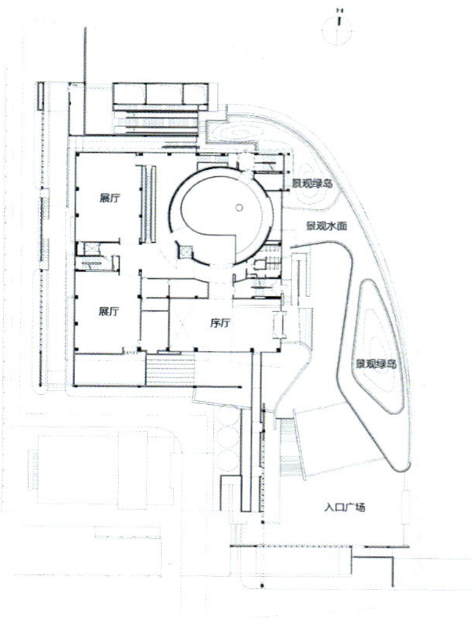

展厅
景观绿岛
景观水面
展厅
序厅
景观绿岛
入口广场

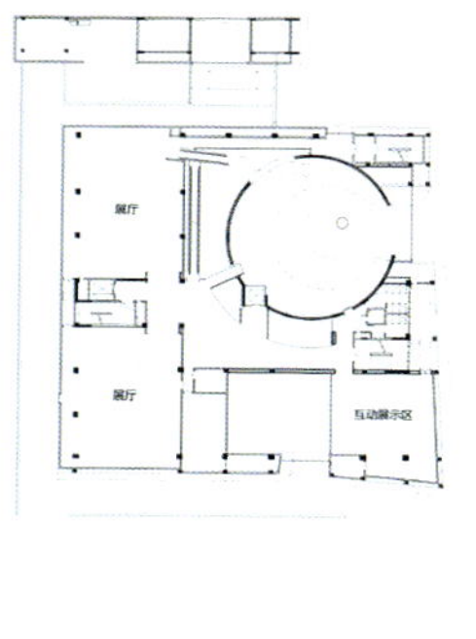

展厅
展厅
互动展示区

李政道图书馆

Li Zhengdao Library

项目位置：上海市
设计时间：2012 年
竣工时间：2018 年
建筑规模：6,500 m^2

Location：Shanghai
Design：2012
Completion：2018
GFA：6,500 m^2

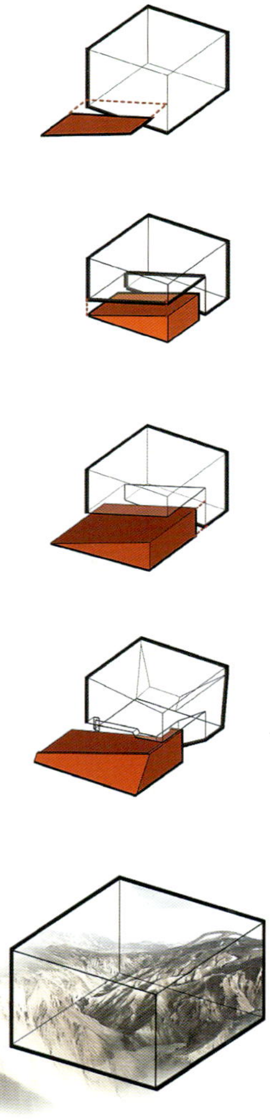

本设计借鉴吴冠中先生《对称乎，未必，且看柳与影》的画作，旨在从建筑上呼应李政道先生的理论研究成果，演绎出由对称与非对称共同构筑的美妙平衡；同时立足于李政道先生 31 岁便获得诺贝尔物理学奖的辉煌成就，取用“诺贝尔山”的意象，旨在激励青年学生和科技工作者坚持不懈，勇攀科学高峰。

设计在对称稳定的体量中引入动感的切割，并最终生成上下叠置的两个单元。位于下方的为三棱体，位于上方的为带切角的方体，其内包含展览、库藏、阅览、办公、科研等功能。方体之内嵌入中庭，为建筑引入光线的同时成为汇聚各种活动的核心区域，为使用者创造层次丰富的交流场所。

Drawing on Wu Guanzhong's *Symmetry ? Not necessarily. Look at Willow and Shadow*, this design aims to echo the theoretical research results of Mr. Li Zhengdao in architecture and deduce the wonderful balance constructed by symmetry and asymmetry. Based on the brilliant achievement of Mr. Li Zhengdao, who won the Nobel Prize in physics at the age of 31, this design takes the image of “Nobel mountain” to encourage young students and scientists to persevere and climb the scientific peak.

The design introduces dynamic cutting into the symmetrical and stable dimension, and finally generates two overlapping units, which are three diamond body at the bottom and square body with cutting angle at the top, including exhibition, library, reading, office, scientific research and other functions. The square body is embedded into the atrium, which introduces sky light into the building and becomes the core area to gather various activities, creating a multi-level communication place for users.

在校园的总体规划方案中，图书馆正处于南北轴线的中心点。为了加强整体校园轴线关系和空间序列，重塑校园礼仪轴线，改扩建后的图书馆将成为校园融合交汇的重要节点；同时为师生之间、同学之间提供交流交往的平台，碰撞出知识的火花。

空间构成以博古架为灵感，赋予建筑"博古汇今"的概念，寓意图书馆亦是存放知识智慧的容器，同时借取博古架形态"精""巧""宜"的特点，融合到设计中，通过对不同功能分区的分析和划分，将功能空间、天井、采光天窗及交通核有机组织与组合。受竹简简牍的启发，抽象的肌理形成了灵动活跃的立面形态，产生了丰富的光影效果，并提升了建筑的文化内涵。

In the overall planning of the campus, Library of Beijing University of Technology is at the center of the north-south axis. In order to strengthen the relationship and spatial sequence of the whole campus axis, and reshape the campus etiquette axis, the rebuilt and expanded library become an important node of campus fusion and intersection, which also provides a platform for communication between teachers and students.

The structure of space is inspired by Antique Shelves, which endows architecture with the concept of "conversant with ancient and modern learning". It implies that the library is also a container for storing knowledge and wisdom, while borrowing the fine and skillful concept of Antique Shelves form. Through the analysis and division of different functional zones, the functional space, patio, day-lighting skylight, and traffic core are organized and combined. Inspired by bamboo slips, we abstract its texture to form a dynamic facade, produce rich light and shadow effects, and enhance the cultural connotation of architecture.

北京工业大学图书馆

Library of Beijing University of Technology

项目位置：北京市
设计时间：2015 年
竣工时间：2019 年
建筑规模：40,463 m^2

Location：Beijing
Design：2015
Completion：2019
GFA：40,463 m^2

建筑基地位于上海交通大学徐家汇校区与城市之间的转角地块。

通过对钱老一生事迹的研究与概括，设计提出了“大地情怀、石破天惊”的概念。

以“方正的石头”寓意钱老心系祖国大地的赤子情怀。方正平直的建筑体量呈向外伸展的态势，简洁有力，抽象表达了钱老贡献一生的戈壁滩上的风蚀岩意象。

以“裂开的石头”之中迸发出东二甲火箭的建筑空间场景，寓意钱老领导的“两弹一星”伟大事业是“石破天惊”般的历史事件。“石头”沿校园道路的延长线从中部裂开，面向城市道路展开最大裂面，“V”字形同高玻璃面透出东二甲导弹实物，“石头”立面的肌理抽象而巧妙地呈现钱老的亲切形象，犹如深情地注视着东二甲导弹。拟像与实物、平面与纵深的交织形成时空对话的场景，突出了人物与其事业的共同主题。

建筑外墙以红灰色为主调，延续了百年徐家汇校区的建筑历史文脉。

The site is located at the corner between Shanghai Jiao Tong University Xujiahui Campus and the city.

The design concept is defined as “love of land and rock split from inside”, which implies the patriotism and the remarkable achievements of Qian Xuesen (also known as Tsien Hsue-Shen) respectively.

The Square Rock symbolizes Qian's patriotic sentiment and high moral character. The square and straight building dimension stretches outward in a concise and powerful manner, which serves as an abstract representation of the wind-eroded rock in the Gobi Desert where Qian devoted himself to.

The Split Rock appears to burst open from an origin core, the building space where the real object of DF-2A missile is on display. This serves as a metaphor that the research and development of the nuclear bomb, missile and artificial satellite directed by Qian are of great historical significance, and makes one feel as if the Square Rock is split from inside by the overwhelming energy of the missile, one of Qian's achievements. The Square Rock is split along the extension line of the campus road and opens up the largest crevice in the facade facing the city. The DF-2A missile is visible to the outside through the glass facade at the V-shaped crevice. The facade fabrics vividly present a smiling portrait of Qian who is gazing at the missile with much emotion. The juxtaposition of simulated image and real object, as well as juxtaposition of plane and depth, establish a dialogue of time and space and highlight the theme of the architecture.

The exterior wall features cervinus tone, inheriting the historical architecture context of the centennial Xujiahui Campus, Shanghai Jiao Tong University.

中山大学科技综合楼

Science and Technology Complex, Sun Yat-sen University

项目位置：广东省广州市
设计时间：2007 年
竣工时间：2012 年
建筑规模：36,000 m^2

Location：Guangzhou, Guangdong
Design：2007
Completion：2012
GFA：36,000 m^2

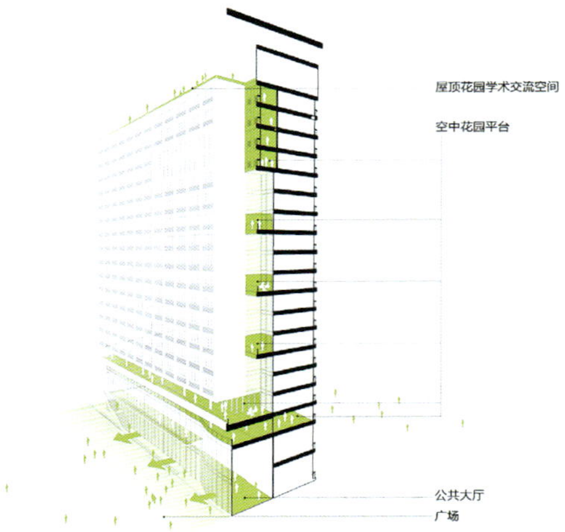

空中架空花园、办公楼层的园林电梯厅，营造了科研创新空间中的非正式交往场所，起到休闲放松、激发创意的作用。空中花园结合绿化环境，在改善环境的同时也加强了建筑物的通风散热效果，以适宜的节能技术实现了生态型办公环境。

The elevated sky garden and the garden-like elevator lobbies on the office floors create the informal social places in the building mainly for scientific research and innovation, which help people relax and inspire creativity. The sky garden, with its green environment, not only improves ventilation and heat dissipation effect in the building, but also builds an ecological office environment with suitable energy efficient technologies.

常州工学院图书馆

Library of Changzhou Institute of Technology

项目位置：江苏省常州市
设计时间：2012 年
竣工时间：2017 年
建筑规模：39,079 m²

Location: Changzhou, Jiangsu
Design: 2012
Completion: 2017
GFA: 39,079 m²

常州工学院图书馆的设计从校园空间、建筑、景观的整合理念出发，综合考虑多方面因素，强调使用功能和交通的合理组织，注重反映江苏省地域气候特征，传承常州古书院文化内涵，致力于创造具有鲜明地域特征和当代建筑空间特色的教育科研人文场所。

图书馆延续校园规划的整合理念，在体量处理上采用了基座、主体与屋顶的三段式，是对传统建筑形态的抽象化处理，并利用现代建筑手法演绎传统建筑元素，从传统角度对龙城书院的概念做出回应，也通过雕塑感强烈的厚重体量凸显了图书馆的标志性。建筑外观色调选择与整体校园协调的“灰”与“红”，在体量突出之余仍能在群体建筑中取得融合。

The design of the library starts from the concept of integration of campus space, architecture and landscape. We consider many factors comprehensively, emphasizing the rational organization of function and transportation. We also focus on reflecting the regional climate characteristics of Jiangsu Province, and inherit the cultural connotation of Changzhou Ancient Academy, aiming to create distinctive regional characteristics as well as educational and humanities places with contemporary architectural space features.

The library continues the concept of integration of campus planning. The three-stage pedestal, main body and roof are used for mass processing. It is an abstract treatment of traditional architectural forms, and uses modern architectural techniques to interpret traditional architectural elements. The concept of the Longcheng Academy is responded, and the library's iconicity is also highlighted by the intense dimension of sculpture. The grey and red of the architectural appearance tone selection and the overall campus coordination can still be achieved in the group building while the dimension is outstanding.

中山大学科技文化交流中心

Science and Culture Exchange Center of Sun Yat-Sen University

项目位置：广东省广州市
设计时间：2007 年
竣工时间：2014 年
建筑规模：38,640 m^2

Location：Guangzhou, Guangdong
Design：2007
Completion：2014
GFA：38,640 m^2

项目集接待、会议、餐饮、办公、住宿于一体，依托北面珠江城市景观和东面中大北湖校园景观，把餐饮设施和住宿设施沿基地东北两侧展开布置。裙楼的架空层花园及塔楼东面的空中花园和退台空中绿化，增加了建筑的观景平台，丰富了公共活动空间，体现了岭南建筑的地域特征。建筑造型稳重大气，既彰显中了大校园形象，也丰富了珠江沿岸的城市界面，成为中大校园的城市名片。

The project integrates reception, conference, catering, office and accommodation into one complex. catering and accommodation facilities unfold along the east and north sides of the site, taking advantage of the cityscape of the Pearl River to the north and the North Lake campus view of the university to the east. Elevated garden of the podium, coupled with the sky garden and terraced sky greening east of the tower, provides additional building landscape platform and varied public spaces while reflecting the regional characteristics of Lingnan buildings. The calm and gracious architectural form of the project, as the icon of the campus, enhances the presence of the university while diversifying the urban interface along the Pearl River.

南京工程学院文体综合馆

Nanjing Institute of Technology Comprehensive Gymnasium

项目位置：江苏省南京市
设计时间：2012 年
竣工时间：2014 年
建筑规模：11,445 m^2

Location: Nanjing, Jiangsu
Design: 2012
Completion: 2014
GFA: 11,445 m^2

南京工程学院文体综合馆的设计概念来自于“更高、更快、更强”的体育奋斗精神，表达了“团结、动感、力量”的设计理念，美好的立意与造型使体育馆具有了独一无二的个性。

综合馆的基本造型可简化为一组流畅灵动的舒展平台翻转成训练馆，其上叠加主馆方正造型的组合。两者一动一静，一灰一白形成鲜明对比，建筑的天际轮廓线充满了灵动的韵律，上下起伏，宛如虬龙，又如波涛，成为凝固的音乐，给人以汹涌澎湃、充满动感与力量的感受。在有限的资金投入下，两个馆采用简单的结构形式和低廉的立面材料，通过体量的对比和巧妙的建筑设计形成完美的组合，动感而不失庄重，流畅而不失凝练，并且表达了团结拼搏的竞技精神。

将主馆和副馆拉开，留出 15 米宽的双层平台通道，既加强了左右两个体育区域的联系，也极大地解决了主副馆的通风采光问题。

The design concept of the gymnasium of Nanjing Institute of Engineering comes from the spirit of “higher, faster and stronger” sports struggle. It expresses the design concept of “unity, movement and strength”. The wonderful idea and shape make the gymnasium unique.

The basic shape of the complex can be simplified as a group of smooth and flexible stretching platforms, which can be turned into a training hall, on which a combination of square shapes of the main hall can be superimposed. The skyline of the building is full of vivid rhythm, ups and downs, likes dragons and waves, which becomes solidified music, giving people a feeling of surging, motion and strength. With limited investment, the two libraries adopt simple structural forms and low-cost facade materials. Through dimension comparison and ingenious architectural design, the two libraries formed by a perfect combination. They are dynamic and solemn, smooth and concise, and express the competitive spirit of unity and struggle.

Opening the main hall and the auxiliary hall and setting aside a double-decked 15-meter-wide platform channel, not only strengthens the connection between the left and right sports areas, but also greatly solves the ventilation and lighting problems of the main and auxiliary halls.

武汉大学文科综合楼

Humanities Institute of Wuhan University

项目位置：湖北省武汉市
设计时间：2014—2017 年
竣工时间：2017 年
建筑规模：21,199 m^2

Location: Wuhan, Hubei
Design: 2014–2017
Completion: 2017
GFA: 21,199 m^2

设计从校园肌理出发，对周边校园空间界面进行了积极的呼应，形成连续的空间界面，将主体建筑放置在西南角，与校园主入口形成对景；南面规划为前广场，形成连续的校园开敞空间。设计顺应原有地形，由南至北逐渐抬高，通过景观处理，形成层次丰富的台地，创造丰富校园空间。造型上汲取了武汉大学特有的文化特征，采用新材料、新构架来展现传统的屋顶形式，通过对材料色调、材质、质感上的精挑细选，最终以现代建筑材料陶棍作为屋面檩条，形成与周边琉璃瓦屋面的呼应和共处。室内中庭空间设计简洁明亮，光线通过天窗进入中庭，在镜面玻璃的反射下，形成层次丰富的室内中庭效果。

The design starts from the campus texture and actively responds to the space interface of the surrounding campus, forming a continuous space interface. The main building is placed in the southwest corner to form the opposite view to the main entrance of the campus; the south is planned as the front square, forming a continuous campus open space. The design conforms to the original topography and is gradually raised from south to north. Through landscape treatment, a rich terrace is formed to create a rich campus space. The shape has absorbed the unique cultural characteristics of Wuhan University, and adopted new materials and new structures to show the traditional roof form. Through careful selection of material tones and textures, the modern building material ceramic rods are finally used as roof purlins, forming an echo and coexistence with the surrounding glazed tile roof. The design of the indoor atrium is simple and bright. The light enters the atrium through the skylight, and under the reflection of the mirror glass, it forms a richly layered indoor atrium effect.

东莞职业技术学院图书馆
Dongguan Vocational and Technical College Library

项目位置：广东省东莞市
设计时间：2007—2008 年
竣工时间：2010 年
建筑规模：21,540 m^2

Location：Dongguan, Guangdong
Design：2007–2008
Completion：2010
GFA：21,540 m^2

图书馆位于湖滨，与教学楼、实训楼与实验楼围绕湖滨布置，总图上形成两个方向的轴线，将教学中心区串联起来。在平面设计上，将阅读空间围绕中庭设置，各种不同类型的阅读空间都包含在一个大空间中。考虑到岭南气候，采用了适应气候的院落式小进深来组织自然通风，架空层与中庭结合在一起加强通风效果。

整体在建筑艺术方面将成为校园中心的一座雕塑，成为一个明显的标志，同时具有庄严朴素的外形。在建筑造型上，强调体型的丰富、体块组合与穿插，强调建筑的力度感。丰富的架空层、中庭和休息空间自然就形成了丰富的建筑造型细节。

The library is located on the lakeside, and is arranged around the lake with the teaching building, the training building and the experimental building. The master plan forms axes in two directions, connecting the teaching center. In the plan design, the reading space is set around the atrium, and various types of it are contained in one large space. Considering the Lingnan climate, it adopts a climate-friendly courtyard in small depth to organize natural ventilation. The overhead layer is combined with the atrium to enhance ventilation.

The whole building will become a sculpture and an obvious symbol in the center of the campus in art, with a solemn and simple shape. The modeling design emphasizes the richness of shape, block combination and intersection, and strength of construction. The overhead layers, atrium and rest space form a rich architectural detail naturally.

福建工程学院新校区图书馆

Library of FuJian University of Technology

项目位置：福建省福州市
设计时间：2008 年
竣工时间：2013 年
建筑规模：40,500 m²

Location：Fuzhou, Fujian
Design：2008
Completion：2013
GFA：40,500 m²

福建工程学院新校区图书馆是校园的中心建筑，也是福州大学城的最高建筑物。项目具有的户外空间和强烈的向心性，考虑到福州温暖湿润的气候特征，通过底层架空层、内院、螺旋上升的空中花园等形成适合当地气候的多层次室外交流场所。建筑裙楼采用半圆形的环廊与地形相呼应，强调了图书馆在校园中的可见性。

The new campus library of Fujian Institute of Technology is the central building of the campus and the tallest building in Fuzhou University Town. The library has a variety of outdoor spaces, and it show centrality in the new campus. For the warm and humid climate characteristics of Fuzhou, we have designed a number of overhead floors, inner courtyards and spiraling sky gardens for the library, which forms a multi-level outdoor communication venue suitable for the local climate. The semi-circular podium echoes the terrain and it makes the library visible in the campus.

临沂大学艺、音、美综合体
Linyi University Art, Music and Fine Complex

项目位置：山东省临沂市
设计时间：2006 年
竣工时间：2011 年
建筑规模：38,665 m²

Location: Linyi, Shandong
Design: 2006
Completion: 2011
GFA: 38,665 m²

"自由与秩序恰当结合是人们的永恒任务。没有自由，就无法发展；没有秩序，就无法持续发展。"

艺、音、美建筑综合体以一条流动的共享步行走廊把艺术中心、音乐学院和美术学院串联起来，脉络清晰有序。建筑整体是在无序中寻求有序，流动的空间形体在有序的背景下，增添了无序的艺术特质。以此为脉络，琴房、室外剧场、室外展场、临水跌落平台等有特殊要求的使用功能和设施配套，均找到了适合自身的合理位置。这种有序与无序辩证结合的布局方式，既延续了艺术特有的气质，也为将来建筑群的按需扩建提供了很有利的前提。多变的空间组合与简单的色彩，让未来的新建部分可以更为容易地融入整体建筑群之中，实现生长型的可持续发展。

"The proper combination of freedom and order is the eternal task of people. Without freedom, there can be no development. Without order, there can be no sustainable development."

The art, music and fine complex connects the art center, the conservatory of music and the academy of fine arts in a flowing and shared walking corridor with a clear and orderly sequence. The building as a whole seeks order in disorder. The flowing spatial form adds disorder to the artistic characteristics under the orderly background. In this context, the piano room, outdoor theater, outdoor exhibition hall, and waterside drop-off platform are all equipped with special required functions and facilities, and have all found their proper positions. This arrangement, which combines order and disorder dialectically, not only continues the unique artistic temperament, but also provides a very favorable premise for the expansion of buildings on demand in the future. The changeable space combination and simple colors make it easier for the future newly-built parts to be integrated into the whole building complex, so as to achieve growth-oriented sustainable development.

吉林大学前卫南区图书信息综合楼设计

Library and Information Complex, Jilin University Qianweinan Campus

项目位置：吉林省长春市
设计时间：2015 年
竣工时间：2018 年
建筑规模：71,800 m^2

Location: Changchun, Jilin
Design: 2015
Completion: 2018
GFA: 71,800 m^2

鼎新楼作为吉林大学核心区域的地标性建筑，体量设计为大致中心对称的布局，中心建筑体量高耸，周边舒展对称。格局延续老校区轴线，突出建筑的统领地位，形成校园高效的空间秩序。建筑面向南北呈半环抱形态，形态借鉴中国古典城市中门楼，强调门户属性，与开放性广场组合形成中心景观特色区，在校园形象上营造出开放、多元、迎接师生的姿态，成为老校区与城市的纽带。

Dingxin Building is a landmark building in the core area of Jilin University. The volume design is roughly a central symmetry. The central building is tall and symmetrical. The pattern continues the axis of the old campus, highlighting the leading position of the building and forming an efficient spatial order on the campus. The building faces the north and south in a semi-enclosed form. The form draws on the gatehouses of the classical Chinese city, emphasizing the portal property, and forming a central landscape characteristic zone with the open square. It creates an open and pluralistic image on the campus, welcoming teacher and student, and becoming the bond with the old campus and the city.

武汉光电国家研究中心

Wuhan National Laboratory for Optoelectronics

项目位置：湖北省武汉市
设计时间：2012 年
竣工时间：2018 年
建筑规模：118,891 m^2

Location：Wuhan, Hubei
Design：2012
Completion：2018
GFA：118,891 m^2

武汉光电国家研究中心位于华中科技大学校内，是全国六大国家研究中心之一，包括专业实验室、科研平台、行车大厅、会议中心等，为一座教学科研综合体。设计以“光电律动，绿韵华中”为设计概念，建筑通过水平与垂直两个维度的叠加、错动，形成一系列有机连续的共享空间，同时，上部连续空间与底层地景空间相互融合，组合出的架空灰空间成为汇聚各种活动的核心场所，建筑力求从多方面达到建筑与科技、艺术、生态、人性的完美交汇融合。

Located in the campus of Huazhong University of Science and Technology, Wuhan National Laboratory for Optoelectronics is one of the six national research centers in China. It includes professional laboratories, research platforms, driving halls, conference centers, etc. It is a teaching and scientific research complex. With the concept of “rhythmic photoelectric and green campus”, the building forms a series of continuous organic shared spaces by overlapping and staggering the horizontal and vertical spaces. At the same time, the upper continuous space and the bottom landscape space merge each other, and the combined gray space becomes the core place for gathering all kinds of activities. The building strives to achieve the perfect integration of architecture and technology, art, ecology and humanization from many aspects.

肇庆学院艺术教学大楼
Zhaoqing College Art Teaching Building

项目位置：广东省肇庆市
设计时间：2011 年
竣工时间：2016 年
建筑规模：19,880 m²

Location: Zhaoqing, Guangdong
Design: 2011
Completion: 2016
GFA: 19,880 m²

一体化的设计理念是艺术教学大楼方案设计的指导思想，同时，艺术教学大楼作为肇庆学院新的标志性建筑，对其艺术文化理念的演绎也是本案的重点。艺术教学大楼外形采用飘带作为母题，墙面上水平向的韵律变化像跳动的音符，暗示其深厚的艺术文化积淀和活跃的空间氛围。在视觉上，不同的色彩塑造着不同的空间氛围，基于不同功能用房的空间性格，赋予不同的空间色彩，形成活跃的视觉焦点，为艺术教学大楼带来充满生机活力的空间感受。

The integrated design concept is the guiding ideology of the design of the art teaching building. At the same time, the art teaching building is the new landmark building of Zhaoqing College. The interpretation of its artistic and cultural concept is also the focus of this case. The shape of the art teaching building uses streamer as the motif, and the horizontal rhythm changes on the wall, like the beating notes, suggesting its deep artistic and cultural connotation and active space atmosphere. Visually, different colors shape different spatial atmospheres. Based on the spatial characteristics of different functional rooms, different spatial colors can be given, which can form an active visual focus and bring a vibrant and dynamic space to the art teaching building.

中国石油大学（北京）实验办公综合楼

China University of Petroleum (Beijing) Experimental Office Complex

项目位置：北京市昌平区
设计时间：2014 年
竣工时间：2015 年
建筑规模：72,000 m^2

Location: Changping, Beijing
Design: 2014
Completion: 2015
GFA: 72,000 m^2

中国石油大学新建综合楼项目为一座集办公、科研于一身的综合性办公楼，其建设地点位于中国石油大学（北京）昌平主校区校园内东南角，东侧为市政亢山路，南侧为中石路。项目用地面积 14,868 ㎡，建设规模为 72,000 ㎡，其中地下部分 15,000 ㎡左右。建筑分为南北两座塔楼、裙楼和地下室；塔楼 15 层，楼高 59.65m; 裙楼 4 层，楼高 16.1m; 地下室 2 层，深度为 8.6m。

The newly-built comprehensive building project of China University of Petroleum is a comprehensive office building integrating office and scientific research. The construction site is located at the southeast corner of the campus of Changping Main Campus of China University of Petroleum (Beijing). The east side is the municipal Kangshan Road and the south side is Zhongshi road. The project covers an area of 14,868 square meters. The construction scale is 72,000 square meters, of which the underground part is about 15,000 square meters. The building is divided into two sections: the north and south towers, the podium and the basement; the tower is 15 stories high and the building is 59.65 meters high; the podium is 4 stories high and the building is 16.1 meters high; the basement is 2 stories and the depth is 8.6 meters.

常州市天合国际学校外国语学校综合体

Changzhou Tianhe International School Foreign Language School Complex

项目位置：江苏省常州市
设计时间：2011 年
竣工时间：2016 年
建筑规模：78,082.83 m^2

Location：Changzhou, Jiangsu
Design：2011
Completion：2016
GFA：78,082.83 m^2

常州高级中学与西侧的常州天合国际学校共享一个地块，如何处理好两个学校的规划布局是该项目的技术难点也是该项目的技术成效。两所学校相依而建，教学设施、课程建设、师资配备等方面最大限度地资源共享是我们设计的出发点，我们采用了整体设计理念，力图形成具有国际学校独特空间形态与传统中学校园空间相结合的整体发展框架，并重点打造“一轴、一带”来组织两个校区的建筑，创造出主次分明、虚实结合、富有韵律的校园建筑空间共享带。

Changzhou High School and Changzhou Tianhe International School on the west side share a plot of land. How to deal with the planning layout of the two schools is the technical difficulty of the project and the technical effect of the project. The two schools are built together, and the maximum of resource sharing in teaching facilities, curriculum construction, and teaching staff is the starting point of our design. We have adopted the overall design concept, and tried to form an overall development framework, which has a unique spatial form of international schools combined with traditional middle school campus space. And we focused on creating “one axis, one belt” to organize the buildings of the two campuses, to create a campus building space sharing belt with distinct primary and secondary, virtual and real, and rhythmic.

湖南女子学院逸夫图书馆

Shaw Library of Hunan Women's University

项目位置：湖南省长沙市
设计时间：2008 年
竣工时间：2010 年
建筑规模：15,431 m^2

Location: Changsha, Hunan
Design: 2008
Completion: 2010
GFA: 15,431 m^2

逸夫图书馆是学校中心的标志性建筑，也是整个校区的学习共享区。图书馆内部采用灵活多变的开放式阅览空间，结构楼板按藏书区荷载采用均布荷载设计，便于模数化的藏书区和阅览区日后互换，使多种学习模式成为可能。外部低层造型采用较大实墙处理，使图书馆不同于一般化的教学建筑立面风格，在略显纷乱的校园环境中营造出平静稳重的气质，成为视觉中心。

图书馆高层区域采用错动的开窗立面，既可将不同方向的城市景色引入图书馆内部，不规则摆放的窗洞也如同随机放置于书架上的一本本书籍，凸显图书馆的建筑属性。

The Shaw Library is a landmark building in the center of the school and a shared learning area throughout the campus. Flexible open reading space is adopted in the library. The structure floor adopts uniform load design according to the load of the book collection area, which facilitates the exchange of modular book collection area and reading area in the future, making it possible for multiple learning modes. The exterior low-rise modeling adopts larger solid wall, which makes the library different from the general teaching building facade style, and creates a calm and stable temperament in the slightly chaotic campus environment and becomes the visual center.

The high-rise area of the library adopts staggered window-opening facade, which can not only bring the urban scenery in different directions into the library, but also display the architectural attributes of the library with irregular window holes which like books placed randomly on the shelves.

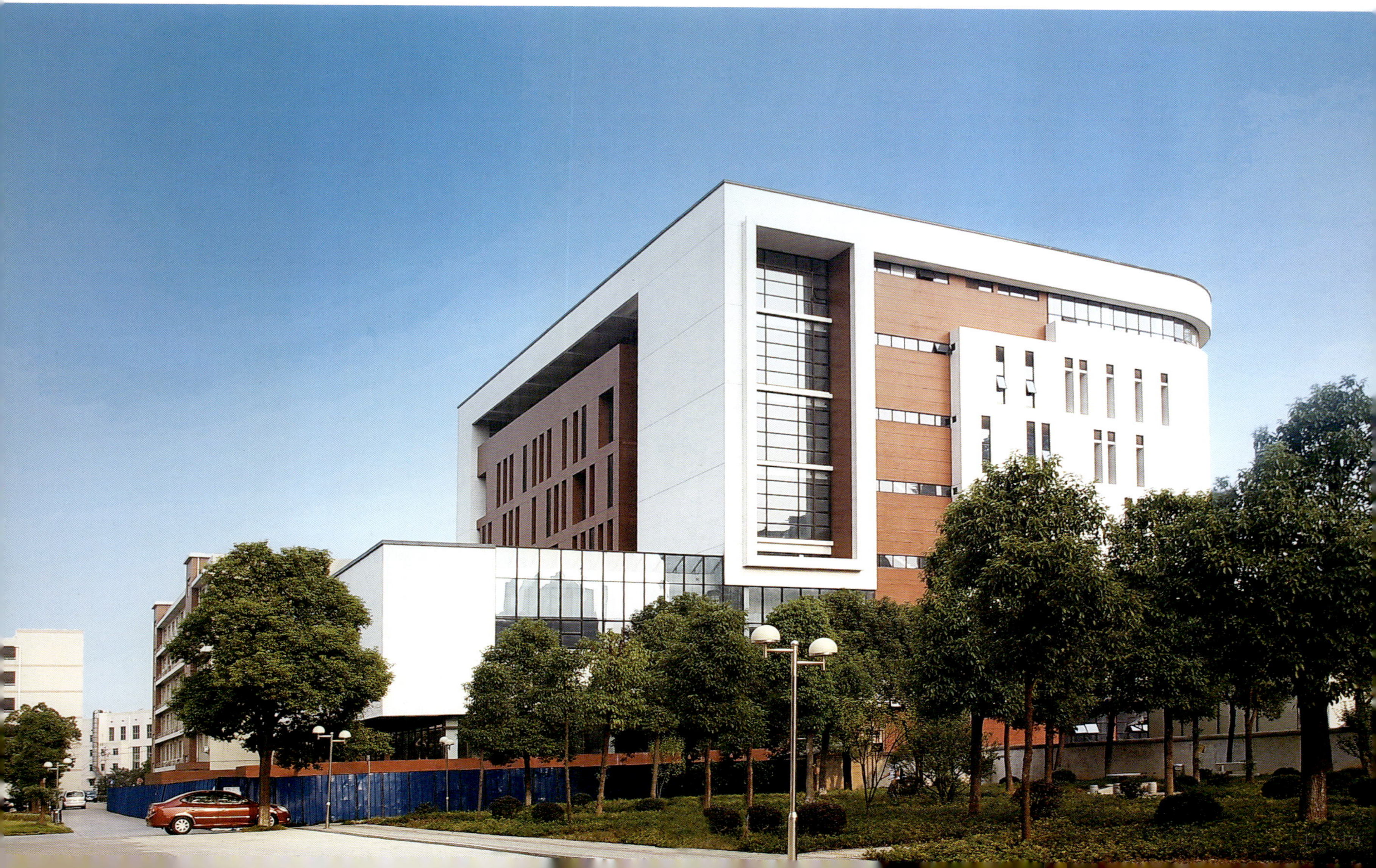

中国地质大学（北京）校园总体规划方案设计

China University of Geosciences (Beijing) Campus Conceptual Master Plan

项目位置：北京市
设计时间：2019 年
建筑规模：706,000 m^2

Location：Beijing
Design：2019
GFA：706,000 m^2

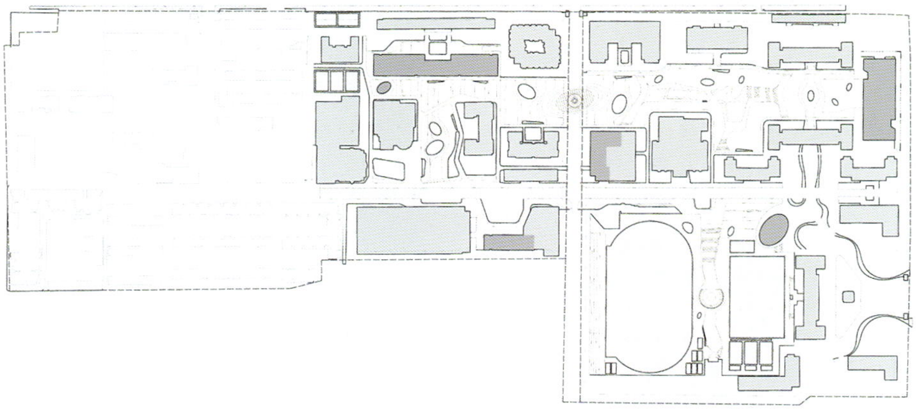

该项目在中国地质大学（北京）现有校区的基础上进行设计，校园现有多个开放空间和轴线序列，但缺乏联系，难以形成连续高效的步行网络，应加强各开放空间之间的联系，形成连续复合的校园室外交往结构。

方案提取现有的片段化的公共空间和轴线关系，拆除部分杂乱分布的老建筑，梳理校园公共空间界面；加强各部分公共空间之间的步行交通联系，形成开合有序的公共空间脉络和整体统一的校园形象；结合校园文化打造多个开放公共空间，塑造有场所记忆的景点。

方案通过充分利用地下空间，发掘校园的最大建设可能性。通过在新建教学楼、宿舍、绿地及广场下方进行大规模的地下空间开发，为校园增加接近 20 万㎡的建筑面积，将中国地质大学（北京）打造成复合研究型、生态节能型、创新示范型校园，引领北京高校改扩建先河。

The project is designed on the existing campus of China University of Geosciences (Beijing). There are many open spaces and axes in the campus, but it is difficult to form a continuous and efficient walking network due to the lack of connection. It is necessary to strengthen the connection between the open spaces and form a continuous and compound outdoor communication area for campus.

The scheme extracts the existing relationship of public space and axis in the campus, demolishes some old disorderly buildings, and reforms the campus public space interface. We strengthen the pedestrian traffic connection among all parts of public space, and form an orderly public space and an overall unified campus image. And we combine with the campus culture to create a number of memory scenic spots in the open space.

By making full use of the underground space, we can explore the maximum construction possibility in the campus. Through the large-scale underground space development under the newly-built teaching buildings, dormitories, parklands and squares, we have nearly 200,000 m^2 construction area more for the campus. At last, the China University of Geosciences (Beijing) will be built into a compound research-oriented, ecological energy-saving and innovative campus. We aim to lead the transformation and expansion construction of universities in Beijing.

广东药科大学云浮校区

Guangdong Pharmaceutical University Yunfu Campus

项目位置：广东省云浮市
设计时间：2017 年
竣工时间：2019 年
建筑规模：408,000 m²

Location: Yunfu, Guangdong
Design: 2017
Completion: 2019
GFA: 408,000 m²

广东药科大学云浮校区选址于云浮市西江新城组团的文教片区内。校园紧邻云都大道、云信大道、康盛路等城市主要干道。附近有即将建成的高品质住宅小区以及城轨站、客运站等重要交通设施。城市生态廊道穿越场地，周边山体、绿地、水体等丰富的景观要素为新校区的建设提供了得天独厚的自然条件。

Yunfu Campus of Guangdong Pharmaceutical University is located in the cultural and educational area of Xijiang New Town, Yunfu. The campus adjoins the main urban trunk roads such as Yundu Avenue, Yunxin Avenue and Kangsheng Road. Premium housing estates and key transportation facilities including an inter-city rail station and a bus passenger station will be completed in the adjacent locations. An urban green corridor running through the site and the surrounding mountains, green space and water bodies provide rich landscaping elements for the new campus.

天津科技大学泰达校区规划设计

Planning and Design for TEDA Campus of Tianjin University of Science & Technology (TUST)

项目位置：天津市
设计时间：2012 年
竣工时间：2014 年
建筑规模：1,004,974 m^2

Location: Tianjin
Design: 2012
Completion: 2014
GFA: 1,004,974 m^2

天津科技大学泰达校区规划设计中既有新建基地，又有旧基地扩建以及功能置换改造，如何处理好众多不同的现状条件以及现有元素，使三个割裂的校区整合为一体，是设计面临的最大挑战。为此，我们提出四个设计理念：一体化的校园格局——通过东西走向的校园人文轴线串连起三大区；核心共享的校园资源——在校园中心设计了中央的核心共享绿化区域，带动各区联动发展；多层次的校园空间——各空间环环相连，营造多层次的人文氛围；可持续的发展规划——综合考虑日后校区发展，连通东西轴线，形成多核心的校园人文礼仪轴线。通过以上四点设计构思，三个校区重新整合为各功能空间一气呵成的整体校园。

The planning and design of TEDA Campus involves new buildings, existing building expansion and functional replacement. Hence, it poses the biggest challenge in design-to properly tackle the existing conditions and elements, combine the three disconnected campuses. Four main concepts are developed in design. First, create an integrated campus layout with the east-west humanistic axis linking up the three main zones. Second, realize sharing of campus resources by designing a shared green zone in the center of the campus to drive interactive development of the zones. Third, form a multi-level campus spaces. The spaces of various levels are all inter-connected logically to form a diverse and highly humanistic atmosphere. Fourth, establish a sustainable development plan with comprehensive considerations to future campus development, for linking up the east axis and the west axis, and forming a humanistic ceremony axis on campus with multiple cores. The above four design concepts help integrate the three separated campuses into a complete one in which functions at spaces appear to be created all in one.

莆田学院新校区修建性详细规划

Detailed Planning of the New Campus of Putian College

项目位置：福建省莆田市
设计时间：2014—2017 年
竣工时间：2017 年
建筑规模：1,980,000 m²

Location：Putian, Fujian
Design：2014-2017
Completion：2017
GFA：1,980,000 m²

项目通过整体性建造与可持续发展，塑造知识经济下最具地域环境特点、时代特征与文化特色的海西地区高端教育平台与人才创新地。基地内外的山水格局为本次规划结构最重要的自然基底，规划形成三大综合组团，生活区邻近相关院系教学区布置，最大化满足师生在步行的距离内学习与生活。项目着力打造“山水相融、三园三带、一廊五轴”的共享空间及校园规划结构，采用了融山聚水、自然入园，组群发展、多核分散，长轴并置、集约共享，人性尺度、人文场所的创新设计理念。

Through holistic construction and sustainable development, the project will shape a high education platform and talent innovation base of the West Side of the Strait Zone Region with the largest extent of geographical environment features, characteristics of the times and cultural characteristics of the knowledge economy. The landscape pattern inside and outside base is the most important natural foundation of the planning structure. Three comprehensive groups are planned to be formed with the living area adjacent to the teaching area of relevant departments, which tries to meet to the great extent that the learning and living of teachers and students are within walking distance. The project strives to create a shared space and campus planning structure of “three mountains and three belts, one corridor and five axes”. Innovative design concepts of blending mountains and gathering waters, nature integrate the garden, group development, multi-core dispersion, long-axis juxtaposition, intensive sharing, humanity scale and humanistic place are adopted.

方太大学
Fotile University

项目位置：浙江省宁波市
设计时间：2017—2018 年
竣工时间：2021 年
建筑规模：63,848 m²

Location：Ningbo, Zhejiang
Design：2017–2018
Completion：2021
GFA：63,848 m²

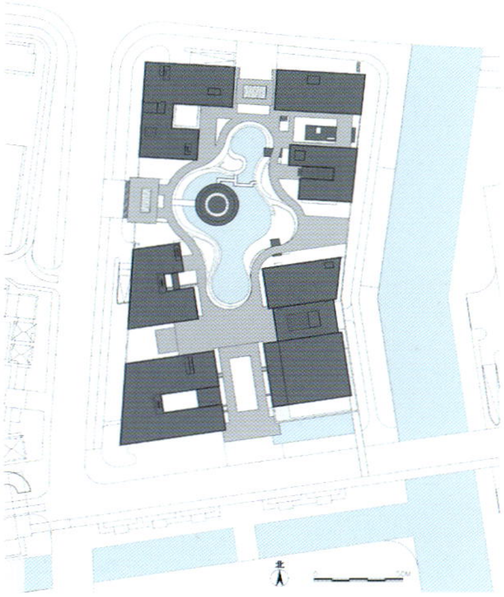

方太大学的设计以儒道思想作为出发点，使企业的儒道文化及儒家管理模式在建筑空间中得以转译。儒道思想强调环境育人的重要性，设计从传统中国园林中获取灵感，通过现代的手法提取传统建筑的院落、庭园、敞廊等空间元素，在符合企业大学使用功能的条件下，塑造具有传统文化内涵的空间层次。

建筑的外轮廓布局方正且遵循秩序，通过隐含的轴线序列隐喻传统文化中的礼仪准则；中央庭院以曲廊组织建筑群落，充分呼应中国传统园林“境贵乎深，不曲不深也”的视觉想象和思想意境，同时形成连续多变的公共交往空间，适应功能需求。

Based on Confucianism and Taoism, which emphasize the importance of the environment's nurture of people, the design proposal of Fotile University interprets the Confucianism spirit of the company and Confucian management model in architectural spaces. Inspired by the traditional Chinese gardens, we utilize various spatial elements in traditional architecture such as courtyards, gardens and corridors to establish a spatial hierarchy with rich cultural connotation within the functional framework of the university.

The buildings feature straight and regular layout, with the sequences of the hidden axes implicating the ritual rules in traditional culture. Building blocks are organized with the zigzag corridors in the central courtyard, fully demonstrating the visual and ideology of “winding path leads to profound scenes” pursued in the traditional Chinese gardens. In doing so, the continuous and varied public spaces are formed to meet the functional needs.

海南大学海甸校区更新改造

Hainan University Haidian Campus

项目位置：海南省海口市
设计时间：2017 年
竣工时间：2024 年（预计）
建筑规模：1,310,000 m^2

Location：Haikou, Hainan
Design：2017
Completion：2024 （expected）
GFA：1,310,000 m^2

设计充分发挥校园区位优势，打开校园南侧的用地界面以改变原有的封闭空间格局，形成校、城、海相融合的开放空间形态。通过尺度宜人的景观开放带串连起东西两湖，强化校园南北轴线形象与空间，串联校园各个空间节点，打造脉络清晰的复合十字形空间架构，形成网络化的整体校园空间。

设计吸取海南当地地域元素，结合绿色建筑改造，创造地域风貌鲜明、生态节能的校园。结合整体空间结构的优化和功能布局的调整，推进校园景观的整体意境塑造，通过特色景区的设计，营造具有浓烈热带风情的美丽校园环境。

The design gives full play to the advantages of campus location, in order to open the land interface on the south side of the campus to change the original closed spatial pattern, forming the open space form of the integration of campus, city and marine phase. It connects the east and west lakes through the landscape open belt with pleasant scale, strengthening the image and space of the north and south axis of the campus and concatenating the various spatial nodes of the campus. In this way, it creates a clear compound cross spatial structure and a whole campus space.

The design absorbs the local elements of Hainan, combining with the transformation of green buildings, so as to create a campus with regional distinct style and ecological energy-saving. In addition, it combined with the optimization of the overall spatial structure and the adjustment of the functional layout, in order to promote the shaping of the overall artistic conception of the campus landscape, and to create a beautiful campus environment with strong tropical customs through the design of the characteristic scenic spots.

前海合作区前湾十单元学校项目
School in Unit 10 of Qianwan, Qianhai New District

项目位置：广东省深圳市
设计时间：2017–2019 年
竣工时间：2020 年
建筑规模：139,240 m^2

Location：Shenzhen, Guangdong
Design：2017–2019
Completion：2020
GFA：139,240 m^2

前海合作区前湾十单元学校项目方案融合环境育人的理念，致力于塑造具有“勇气、荣誉、谦和、互助”的哈罗价值观和前海精神的教学环境。规划设计打破用地限制，让国际学校与民办学校共享场地，实现统一整体的校园空间格局。

建筑体量向西南面河道公园层层跌落，与城市景观形成良好关系。校园空间向河道公园敞开，将滨河公园景观引入场地，主要的教学生活空间均可共享滨河景色。

建筑空间传承了岭南建筑的空间特征，采用庭院式的布局，以及骑楼架空、通风廊道、多层次的架空及遮阳空间等岭南元素，结合立体化的绿化策略，营造了多层次的岭南园林体验。

School in Unit 10 of Qianwan, Qianhai New District integrates the concept of environmental education and is committed to shape a teaching environment with Qianhai spirit and Harrow values of courage, honour, humility and fellowship. The planning and design break the land restriction and allow international schools and private schools to share the space, so as to achieve a unified overall campus spatial pattern.

The dimension of the building falls to the southwest of river park, forming a good relationship with the urban landscape. The campus space is open to the river park, where the landscape of the riverside park is introduced into the venue, so the main teaching and living spaces can share the riverside scenery.

The architectural space inherits the spatial characteristics of Lingnan architecture, and adopts the layout of courtyard, the elevated building, ventilation corridor, multi-level overhead and sunshade space and other Lingnan elements. It combines the three-dimensional greening strategy to create a multi-level Lingnan garden experience.

ILA Qianhai School

广东以色列理工学院

Guangdong Israel Institute of Technology

项目位置：广东省汕头市
设计时间：2015 年
竣工时间：2017 年
建筑规模：90,500 m²

Location: Shantou, Guangdong
Design: 2015
Completion: 2017
GFA: 90,500 m²

新型的国际性合作大学应当是传承校园历史文化、独具空间特色的个性化校园。通过研究分析，我们总结出了拱券与叠涩、光影和序列、浅色立面等以色列现代建筑的特有风格，并将其作为融入地域文化的承接点。

校区地形极具特色，用地狭长，场地较为平坦。设计将主要功能建筑沿城市干道形成条状布置，宽度约 20m，满足教学实验的功能平面要求。建筑造型呈起伏的山形，既与北侧的桑浦山互相呼应形成山谷之势，新颖的山形建筑造型又突出了以色列理工学院作为世界一流大学所具有的勇于创新的校园风格。建筑中部以学生食堂将教学区与生活区区分，既创造了建筑的节奏感，又避免了对已建成的汕头大学交流中心的视线遮挡，与汕头大学融为一体。建筑通过架空、平台、连廊、空中花园将绿色引入，构成了立体、活跃的研究型校园氛围和空间感受。低碳节能技术的引入以及一系列适宜生态技术的采用，将新校区打造成示范性的可持续发展的绿色生态校园。

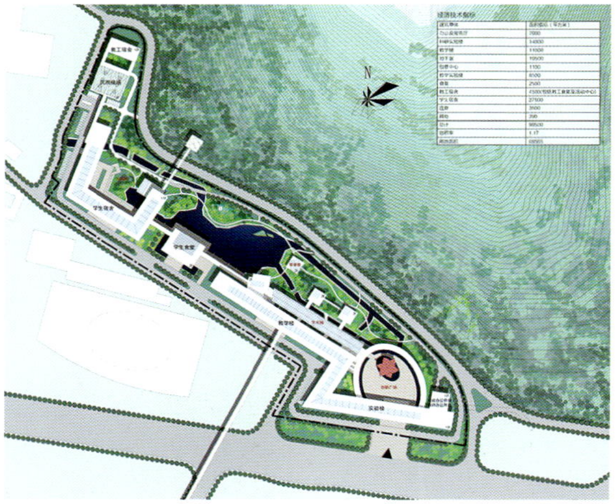

A new type of international cooperative university must be a unique one that not only carries on the heritage of the original university but also has its own characteristics in space. Following studies and analysis, we decided to incorporate the unique elements of modern Israeli architecture into the design, such as arches, shadow and sequence, and light-colored facade and make them integrated with the local culture.

The long and narrow site is quite flat, and major functional buildings are arranged along the municipal road in a belt layout of about 20 meters wide to ensure the functional requirements of teaching and experiments. The building form is an imitation of the undulating mountain, which not only echoes to the Sangpu Mountain. on the north but also demonstrates a unique and innovative campus style that Israel Institute of Technology as a world-class university. In the center of the site, a student canteen separates the teaching and lab area from the living area, creating a rhythm of architecture, avoiding blocking views to the Exchange Center of Shantou University, and making the site fully integrated with the campus of the university. With elevated ground floors, platforms, corridors and sky gardens, the greenery is introduced into the site, fostering a dynamic campus atmosphere for campus research ambiance and spatial experience. The introduction of low-carbon energy-saving technologies and a number of ecological technologies makes the new campus an example of green ecological campus that supports sustainable development.

柳州市委党校搬迁工程

Party School of CPC Liuzhou City Committe

项目位置：广西壮族自治区柳州市
设计时间：2017 年
竣工时间：2019 年
建筑规模：59,960 m²

Location：Liuzhou, Guangxi
Design：2017
Completion：2019
GFA：59,960 m²

以“中国特色的行政院校”为设计理念，最大限度利用原有水体及地形地貌，打造富韵味的中式园林区。以地域化、园林化、生态化”为设计原则，留出大片的绿化区和水面，为创造绿色生态校园创造了良好的基础。

With the design concept of “administrative colleges with Chinese characteristics”, we will make full use of the original water and topography to create a rich Chinese garden. With the design principle of “localization, gardening and ecologicalization”, the large greening area and water area surface have created a good foundation for creating a green ecological campus.

南京航空航天大学天目湖校区

Tianmuhu Campus of Nanjing University of Aeronautics and Astronautics

项目位置：江苏省溧阳市
设计时间：2018 年
竣工时间：2019 年
建筑规模：621,400 m^2

Location：Liyang, Jiangsu
Design：2018
Completion：2019
GFA：621,400 m^2

校园与城市相互融合，形成校城共享的城市校园，实现校园内部学科之间的共享联通，形成整体的校园公共空间。设计延续江南水乡文脉，营造校园的人文特色，实现“环境育人”的教育理念，营造绿色、生态、可持续的海绵校园。

The campus integrate with the city to form the urban campus shared by the school and city. Realize the sharing and communication between the internal disciplines of the campus to form an overall campus public space. The design Continue the cultural context of Jiangnan, create the humanistic characteristics of the campus, and realize the educational concept of “environmental education”. Create a green, ecological and sustainable sponge campus.

04 会展建筑
EXHIBITION ARCHITECTURE

青岛国际会议中心	Qingdao International Conference Center（QICC）
长沙国际会议中心	Changsha International Conference Center
广州市城市规划展览中心	Guangzhou Urban Planning Exhibition Center
广州南沙明珠湾开发展览中心工程设计	Guangzhou Nansha Pearl Bay Development Exhibition Center Project Design
西安曲江会展中心新馆	Xi'an Qujiang New International Exhibition and Convention Center
福建晋江国际会展中心	Fujian Jinjiang International Exhibition and Convention Center
保利 C+ 国际博览中心	Poly C+ International Expo Center
常德市柳叶湖国际会展中心	Changde Liuye Lake International Exhibition and Convention Center

青岛国际会议中心

Qingdao International Conference Center（QICC）

项目位置：山东省青岛市
设计时间：2017 年
竣工时间：2018 年
建筑规模：54,000m²

Location：Qingdao, Shandong
Design：2017
Completion：2018
GFA：54,000 m²

20

青岛国际会议中心是 2018 年上海合作组织峰会的主会场。项目选址于青岛奥帆基地，场地依山面海，风景优美，青岛山、海、城、港的精华汇聚于此，2008 年第 29 届奥运会和第 13 届残奥会的帆船比赛也在此举行。

该项目建成后运营良好，广受全球媒体关注，中央电视台、凤凰卫视等知名媒体做了专题报道，山东省委、青岛市委专程给华南理工大学建筑设计研究院发来了感谢信，获得了社会各界的广泛好评。

Qingdao International Conference Center (QICC), as the main venue of the 2018 Shanghai Cooperation Organization Summit, is located in Qingdao Olympic Sail Base, with beautiful scenery on the mountains and seas. The essence of Qingdao mountain, sea, cities and harbors are gathered here. Sailing competitions of the 29th Olympic Games and the 13th Paralympic Games in 2008 were also held here.

After the completion of the project, the project has been run well and concerned by the global media widely. CCTV, Phoenix Satellite TV and other well-known media have made special reports. CPC Shandong Provincial Committee and CPC Qingdao Municipal Committee have sent thank-you letters to the Architectural Design and Research Institute of South China University of Technology, which has won extensive praise from public.

长沙国际会议中心
Changsha International Conference Center

项目位置：湖南省长沙市
设计时间：2018 年
竣工时间：2020 年
建筑规模：302,000 m^2

Location：Changsha, Hunan
Design：2018
Completion：2020
GFA：302,000 m^2

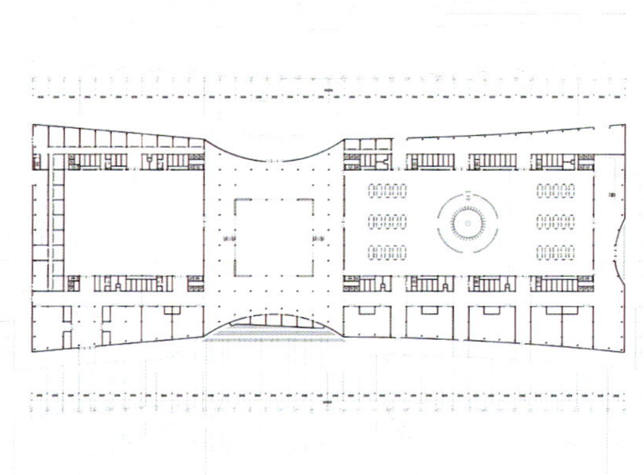

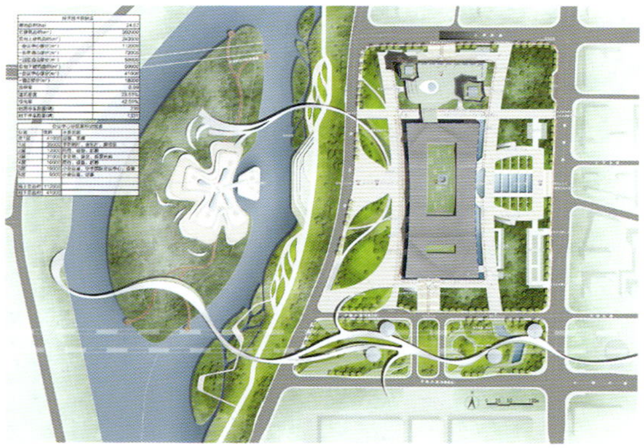

长沙国际会议中心综合体项目主要功能是承办国际中高端会议及配套酒店服务等。项目位于长沙高铁会展新城会展片区浏阳河东岸，与长沙火车南站隔河相望，南侧为已建成的长沙国际会展中心。项目是长沙向世界展示的新舞台、新名片，作为会展片区的先期开发区域，将成为长沙市、湖南省乃至中部地区或全国的城市产业、城市形象的新地标。

As a comprehensive project to undertake international high conference and supporting hotel services, Changsha International Conference Center is located on the east bank of Liuyang River in the exhibition area of Changsha High-speed Railway Convention and Exhibition City, across the river from Changsha Railway South Station, and on the north side of Changsha International Convention and Exhibition Center. The project is a new stage and new business card for Changsha to show to the world. As an early development area of the exhibition area, it will serve as a new landmark of industry and image of Changsha City, Hunan Province, the central region, and even China.

广州市城市规划展览中心

Guangzhou Urban Planning Exhibition Center

项目位置：广东省广州市
设计时间：2013 年
竣工时间：2017 年
建筑规模：84,635 m^2

Location: Guangzhou, Guangdong
Design: 2013
Completion: 2017
GFA: 84,635 m^2

展览中心位于广州市白云山西麓，是白云新城的文化中心。我们通过营造序列化的空间场所，弱化建筑边界，将景观引入，向城市打开，实现建筑、山、水、城一体。架空庭院借景白云山，下方水院与东侧河涌叠水相融，景观廊桥飞架水面之上，形成独具岭南特色的入口空间；空中展厅能环视四周，新市民公园一览无遗；白云步道的大面积透明界面，拾级而上把云山景致尽收眼底。展览馆向南、东两面打开的城市灰空间强调与城市地下空间系统连通，为市民提供一个开放的、可自由进入的全天候室外展场和富有活力的城市客厅。展览馆外表面覆盖两种细密的表皮肌理——模仿岭南民居外墙质感的深色镂空金属板、具有传统韵味的暖色陶棍百叶，用时代性的建筑语言展现城市历史信息。

The exhibition center is located at the west foot of Baiyun Mountain, which is the cultural center of Baiyun New Town. We created a series of spaces to weaken the building boundaries, introducing a new landscape to the city and realizing the integration of architectures, mountains, water and the urban areas. The aerial courtyard takes advantage of the scenery from Baiyun Mountain, and the water courtyard beneath overlaps with the water flowing from the east side. The landscape bridge flies across the water surface, forming an entrance space marked by unique Lingnan characteristics. The air exhibition hall can look around, and the New Citizen Park has an unobstructed view; the large transparent interface of the Baiyun Trail provides a panoramic view of Baiyun Mountain. The gray space of the exhibition hall opening to the south and east,underlines its connectivity with the underground space of the city provides an open and freely accessible 24-hour outdoor exhibition area and a vibrant urban living room for the public. The exhibition hall is covered with two fine skin textures–the dark hollowed-out metal plate that imitates the exterior wall texture of Lingnan folk houses and the warm colored ceramic stick shutter exuding traditional charm. The historical information of the city is well conveyed by the architectural language.

广州南沙明珠湾开发展览中心工程设计
Guangzhou Nansha Pearl Bay Development Exhibition Center Project Design

项目位置：广东省广州市
设计时间：2014 年
竣工时间：2015 年
建筑规模：11,728 m²

Location：Guangzhou, Guangdong
Design：2014
Completion：2015
GFA：11,728 m²

广州南沙明珠湾开发展览中心位于南沙区灵山岛凤凰一桥的东侧，建筑面积约 1.17 万㎡，主要功能区包括明珠湾规划建设展览馆（裙楼）、南沙城市建设投资开发建设指挥部现场办公场所（6层塔楼）两部分。展示馆面临江面，游客可在游览展览馆时欣赏开阔的江景，南侧办公区能远眺明珠湾总部办公区和蕉门水道。

广州南沙明珠湾开发展览中心以地景建筑这一形式为切入点，围绕着“桑基鱼塘”这一再现地域文化印记的母体展开，并以此为符号，同时切分上人屋顶和室外环境平面，使建筑和景观在环境中融为一体，并在立面构造、绿植屋面设计方面引入绿建节能技术。它不仅是一座集规划展览和建设指挥部于一体的复合绿色建筑，还是一座开放的城市公园，更是南沙的地域文化印记和大地肌理在新区的延续。

“鱼塘印记，隐而成景”是南沙明珠湾开发展览中心建筑设计的基本理念。建筑不凸显自己的力量，而是在环境中隐匿且化作景致，有望成为南沙新区的首开建筑。

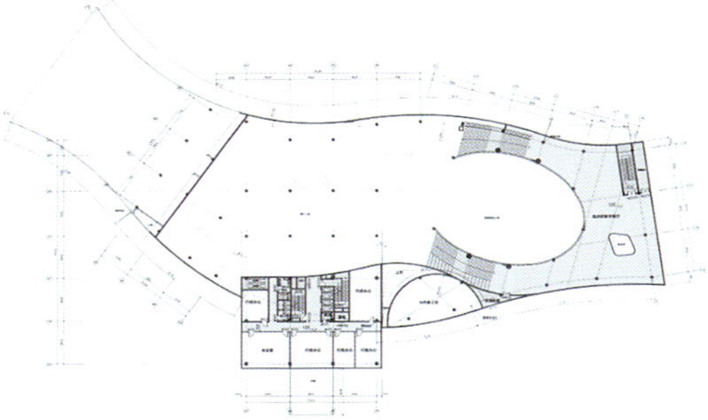

Guangzhou Nansha Development Exhibition Center is located in the Nansha district, Lingshan island, on the east side of phoenix bridge, which covers an area of about 11,728 m². The building consists of two main parts, the one is planning exhibition hall (skirt building), the other is Nansha city construction investment development and construction headquarters site offices (6 floors tower). The exhibition hall faces the river. Visitors can enjoy the wide river view when visiting the exhibition hall. Meanwhile, people also could enjoy the sight of Pearl Bay headquarters and Jiao men waterway when staying in the south office area.

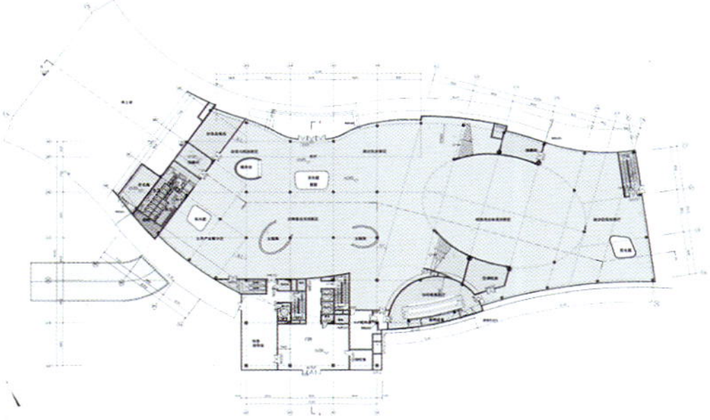

Based on the form of landscaping construction, we choose “mulberry dike fish pond” as a local culture symbol of Guangzhou Nansha Pearl Bay Development Exhibition Center. We take advantage of people accessible roofs design to integrate the architecture and landscape. And energy-saving technologies are adopted in the design of facade and green roof. It is not only a green complex which combines the program of exhibition hall and office, but also a kind of continuation of Nansha regional culture.

“fish pond marks, hidden into the landscape” is the basic concept of the architectural design of Nansha Pearl Bay development and exhibition center. The design minimized the construction impact to the environment by hiding the building in the environment and turning it into a part of landscape. And it will be one of the first buildings in Nansha New District.

西安曲江会展中心新馆

Xi'an Qujiang New International Exhibition and Convention Center

项目位置：陕西省西安市
合作单位：gmp 建筑事务所
设计时间：2006 年
竣工时间：2007 年（一期）
建筑规模：151,866 m^2

Location：Xi'an, Shaanxi
Partner：gmp Architects
Design：2006
Completion：2007 (phase 1 of 3)
GFA：151,866 m^2

项目位于西安历史老城以南，在城市南北轴线长安南路、三环路一侧。在不破坏原有前广场的基础上，一条南北走向的交通长廊将现有展厅与新会展中心连接起来，保持了原有展馆与环境的关系，又很好地组织了与新展馆的联系。7 个标准展厅沿一条功能主轴中廊线性排列，南北展厅错开排列，这样的布展有利于形成连续顺畅的参观流线，也可高效率地组织交通，保障私人及公共交通的畅通。

The project is located on Chang'an South Road and Third Ring Road. On the basis of not destroying the original square, a north-south traffic corridor connects the existing exhibition hall with the new exhibition center, maintains the relationship between the original exhibition hall and the environment, and organizes the new exhibition hall very well. Seven standard exhibition halls are linearly arranged along a functional main corridor. This arrangement is conducive for continuous and smooth circulation, and can also organize traffic efficiently and ensure the smooth flow of private and public transportation.

福建晋江国际会展中心

Fujian Jinjiang International Exhibition and Convention Center

项目位置：福建省晋江市
设计时间：2016 年
竣工时间：2020 年
建筑规模：177,800 m^2

Location：Jinjiang, Fujian
Design：2016
Completion：2020
GFA：177,800m^2

本方案位于博览区的中轴上，因此我们希望以一条绿色景观轴穿过场地来串联整个片区。建筑保持城市轴线的连续性，同时巧妙地利用其作为连接会展各个功能区的主要公共空间。中央生态走廊顶棚造型从“丝绸”和“海洋”意象出发，形成了光滑柔软的曲面造型，营造出气候舒适性和场地标志性。

The project is located on the central axis of the Jinjiang Expo area, so we hope to connect the entire area with a green landscape axis through the site. The center maintains the continuity, of the city's axis while subtly utilizing it as the main public space connecting the various functional areas of the exhibition. The roof shape of the ancient central ecological corridor starts from the image of “silk” and “ocean”, forming a smooth and soft curved surface, providing climate comfort and site landmark.

保利 C+ 国际博览中心
Poly C+ International Expo Center

项目位置：海南省三亚市
设计时间：2019 年
竣工时间：2022 年
建筑规模：54,003 m^2

Location: Sanya, Hainan
Design: 2019
Completion: 2022
GFA: 54,003 m^2

本项目由展览中心和会议中心两座建筑单体组成，设计坚持实用、经济、美观的原则，力求创作一座与时俱进、高调雅致的展览建筑。

本案位于三亚海棠湾旅游度假区内，地块中央主轴线开敞成为大尺度的线性公共空间，承接了城市的轴线关系，是人们汇集、交流的核心室外空间，也是对外形象的重要名片之一。设计始终贯彻“实用、经济、美观”的指导方针，在各个方面、各个环节严格按照国家强制标准、规范。

The project consists of two individual buildings, the exhibition center and the conference center. The design adheres to the principles of practicality, economy and beauty, striving to create an elegant expo center that keeps pace with the times.

The project is located in Haitang Bay Tourist Resort, Sanya. The central main axis of the land is open and becomes a large-scale linear public space which serves as the axis of the city. It is the core outdoor space where people gather and communicate as well as one of the important external images. The design adheres to the guidelines of “practicality, economy and beauty” and strictly complies with national mandatory standards and regulations in all aspects.

常德市柳叶湖国际会展中心

Changde Liuye Lake International Exhibition and Convention Center

项目位置：湖南省常德市
设计时间：2017 年
竣工时间：2022 年（预计）
建筑规模：120,000 m^2

Location: Changde, Hunan
Design: 2017
Completion: 2022 (expected)
GFA: 120,000 m^2

柳叶湖国际会展中心位于常德市武陵区柳叶湖畔。作为城市空间的重要节点，本方案以“柳湖玉滟、水鸣石涧”为意，各栋建筑如宝石一般镶嵌在柳叶湖畔，刚劲夺目，熠熠生辉。中廊则以“流水”抽象成流畅的曲面造型，形成串联各个功能模块的核心公共空间，呼应城市历史文化。方案以宝石切割为母题来塑造刚劲锐利的建筑形象，试图将常德特有的水文化的设计观念融入建筑中，使会展空间与柳叶湖景融为一体。

Liuye Lake International Exhibition and Convention Center is located on Liuye Lake in Wuling, Changde. As an important node of urban space, the design has the image of “sparkling lake and flowing mountain streams”. The buildings are set in the shore of Liuye Lake like gems, shining and dazzling. The smooth curved surface of the center corridor is the abstract of flowing water, which forms a core public space for each functional module in series, and responds to the historical culture of the city. The design uses gem cutting as the motif to create a strong and sharp image, trying to combine the buildings with local water culture and integrate the exhibition space into the view of Liuye Lake.

05 体育建筑
SPORTS ARCHITECTURE

深圳市宝安体育场	Shenzhen Bao'an Stadium
淮安市体育中心	Huai'an Sports Center
江门市滨江体育中心	Jiangmen Sports Center
2010 年亚运会武术馆（南沙体育馆）	Wushu Competition Hall of 2010 Asian Games（Nansha Gymnasium）
2010 年亚运柔道 / 摔跤馆（广州大学城华工体育馆）	Judo & Wrestling Gymnasium of 2010 Asian Games (Gymnasium of SCUT, Guangzhou University Town)
2010 年亚运会省属场馆游泳跳水馆	Swimming Center of 2010 Asian Games
天津科技大学体育馆	The Gymnasium in Tianjin University of Science and Technology
河海大学综合体育馆	Hohai University Comprehensive Gymnasium
东莞市长安镇体育公园体育馆	Dongguan Chang'an Gymnasium in Chang'an Sports Park
武汉大学卓尔体育馆	Gymnasium of Wuhan University
柳州市李宁体育馆	Li Ning Stadium in Liuzhou
常州工学院体育馆	Gymnasium of Changzhou Institute of Technology
内蒙古行政学院文体活动中心	Sports Center of Inner Mongolia Administration College

深圳市宝安体育场
Shenzhen Bao'an Stadium

项目位置：广东省深圳市
合作单位：gmp 建筑事务所
设计时间：2008 年
竣工时间：2011 年
建筑规模：97,712 m^2

Location：Shenzhen, Guangdong
Partner：gmp Architects
Design：2008
Completion：2011
GFA：97,712 m^2

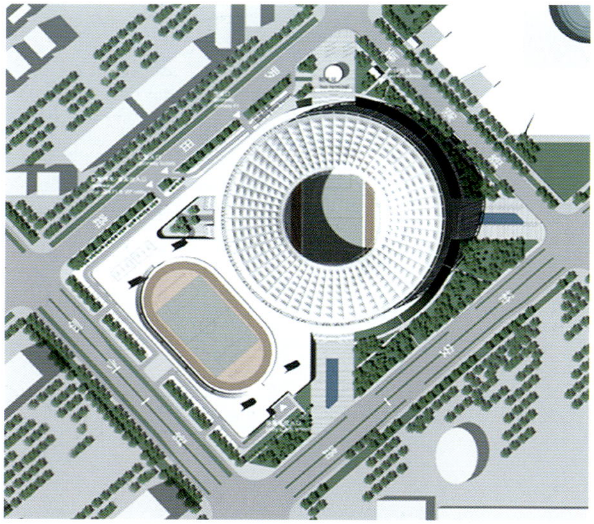

深圳宝安体育场在总平面规划上顺应宝安区中央体育轴线的设计概念，主体设计为一个正圆，以配合周边道路和建筑的不同朝向。体育场设计以深圳特有的亚热带竹林作为建筑设计的灵感。体育场屋盖采用轻盈的膜结构，使整个屋盖看上去犹如一朵巨大的浮云升腾于竹林之上，塑造了独特的建筑外观，呼应了2011年深圳大运会这一欢乐盛事的主题。功能设计充分考虑赛时与赛后的需求，实现体育场在长远时期内的可持续发展。体育场设计造型优雅，功能合理，达到国际一流、国内领先水平。

Shenzhen Bao'an Stadium responds to the concept of central sports axis of Bao'an in master plan. The body has been designed to be a circle to coordinate with the different orientations of surrounding roads and buildings. The stadium is inspired by the unique semitropic bamboo forest in Shenzhen. The light membrane structure of the stadium makes the entire roof look like a huge cloud rising above the bamboo forest, creating a unique appearance that echoes the theme of the 2011 World University Games held in Shenzhen. Functional design fully considers the needs at and after the game, achieving long-term sustainable development of the stadium. The design is elegant and functional, and it has achieved world-class leading level.

淮安市体育中心
Huai'an Sports Center

项目位置：江苏省淮安市
设计时间：2009 年
竣工时间：2014 年
建筑规模：164,411 m^2

Location：Huaian, Jiangsu
Design：2009
Completion：2014
GFA：164,411 m^2

淮安市体育中心是为 2014 年江苏省第十八届省运会而建造的。我们将该区域定位为：以体育产业为主导，以运河文化为纽带，集竞技比赛、休闲健身、商务会展、创意产业、生态体验、文化娱乐、体育旅游、健康宜居等多功能于一体的运动型城市活力中心。

淮安在中国古运河文化时代是一个重要的枢纽。在建筑形象构思上，体育场抽象了“运河文化”中最重要的运输工具“船”的一些形象要素，如拉索、帆等；体育馆、游泳馆综合体则取“水波”之意。将这些要素融合于体育建筑设计中，希望能对这个城市的历史记忆有所回应。

Huai'an city sports center is build for the eighteenth Provincial Games 2014 held in Huai'an City, Jiangsu Province. Our design concept of this area is a sports vibrant city center, which major in sports facilities, canal culture as a link, combining athletic competition, leisure and fitness, commerce exhibition, creative industries, ecological experience, culture and entertainment, sports tourism, health and livable living and so on.

In ancient canal culture times of China, Huai'an city is an important hub city. About the architectural image design, for the stadium, we abstract some image elements of the most important means of transport “ship”, such as cable and sail; for gymnasium and natatorium, we take a reference to “wave”. We combine these elements with the sports architecture design as a memory of the city's history.

江门市滨江体育中心
Jiangmen Sports Center

项目位置：广东省江门市
设计时间：2013 年
竣工时间：2017 年
建筑规模：212,000 m^2

Location: Jiangmen, Guangdong
Design: 2013
Completion: 2017
GFA: 212,000 m^2

江门市滨江体育中心位于江门市北部滨江新城启动区的核心位置。整个体育中心用地范围紧凑，但由于建设规模庞大，如果按照以往传统的设计范式，整个体育中心的室外空间将非常拥挤。建筑师从城市的角度出发，创造性地将建筑紧凑地布置在用地的外围，使建筑更加贴近城市的同时，也让出更多的空间作为城市公共空间和公共绿地。

体育中心采用园林化的布局，体育场与游泳馆区位于场地的南侧，方便周边住宅区人群使用。体育馆与会展区位于场地的北侧，与用地北侧商业区形成呼应，会展和体育馆的两种功能实现相互转换，在日常使用中成为相互的补充。水景公园区位于场地的中间，延续了东侧外围城市的中轴景观带，并沟通联系西侧天沙河景观带。运动场地区位于体育场北侧，是中心景观公园开敞空间的延续。城市广场区及停车区位于建筑的东西两侧，有利于人流集散与车流组织。

Jiangmen Sports Center is located at the core of the Binjiang New Town start-up area in northern Jiangmen. Given the compact site area but massive building volume, the conventional design paradigm will lead to insufficient outdoor spaces. To address the challenge, the architect considers the project from a larger urban context and centralizes the buildings at the periphery of the site, drawing the buildings closer to the city and sparing more area for public spaces and greening.

A garden-style layout is adopted for the sports center. The stadium and natatorium area is placed in the south of the site to facilitate access of citizens from the surrounding residential areas, while the stadium and exhibition area is in the north to respond to the commercial area on its north.Two functions–the gymnasium and the convention and exhibition center are mutually switchable and complementary in daily use. The waterscape park area in the center extends the central urban landscape axis on the east and connects the Tiansha River landscape belt on the west. The sports field on the north of the stadium extends the open space in the central park. The urban square and parking area on the east and west of the buildings facilitate smooth pedestrian and vehicular circulations.

1- 观众看台
2- 门厅
3- 餐厅
4- 管理办公区
5- 二层平台
6- 天井
7- 贵宾门厅

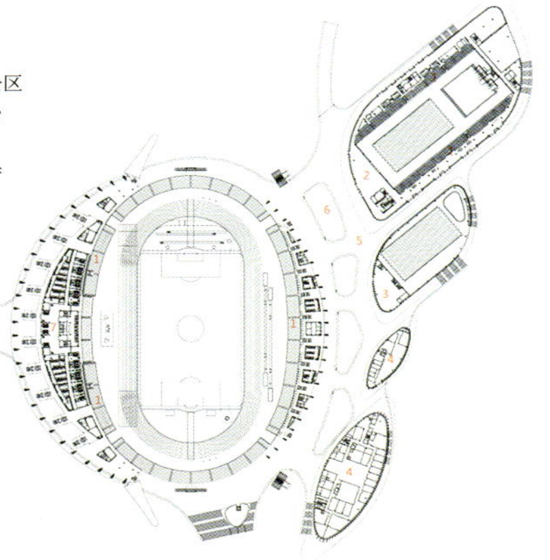

1- 比赛池
2- 跳水池
3- 训练池
4- 门厅
5- 餐厅
6- 商铺
7- 商业步行街
8- 田径运动场
9- 贵宾门厅
10- 运动员区
11- 赛事组织
12- 新闻媒体区
13- 安保
14- 管理用房

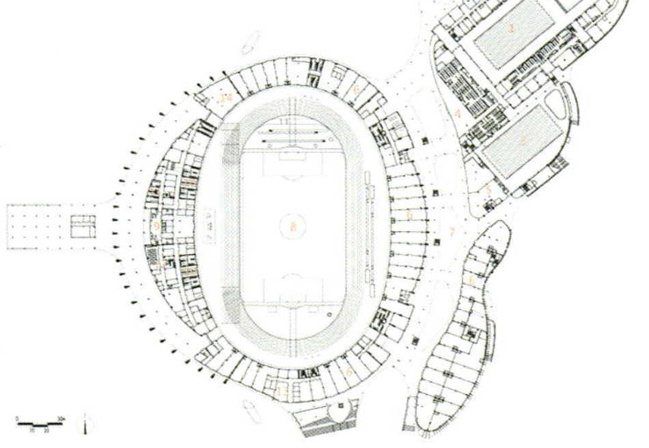

2010 年亚运会武术馆（南沙体育馆）

Wushu Competition Hall of 2010 Asian Games（Nansha Gymnasium）

项目位置：广东省广州市
设计时间：2007 年
竣工时间：2010 年
建筑规模：30,236 m^2

Location: Guangzhou, Guangdong
Design: 2007
Completion: 2010
GFA: 30,236 m^2

南沙体育馆作为 2010 年广州亚运会武术及卡巴迪比赛馆，总建筑面积超 3 万m²，高 29m，总座席数为 8000 多座，其中固定坐席 6000 座，活动坐席 2000 座。

建筑平面形态与太极图的巧合，给了我们巧妙利用“武术”概念的机会。设计中，将组成体育馆外壳的九个曲面单元片片层叠，并分为南北两组，以比赛大厅圆心为中心呈螺旋放射状展开。这样处理的目的是将单一的建筑体量一分为二，并以一种富有动感的方式将两者紧密联系，运用近似太极阴阳图的构成方式，隐喻中国武术的最高境界——“阴阳俱合，天人合一”。

此外，结合广东地区独特的地域文化——海洋文化的特征，我们又借鉴了富有肌理变化的“海螺”外壳作为造型设计的意向，通过屋面金属片片层叠的处理及颜色深浅的变化，力求创造出一种蕴含了地域文化特征的建筑形态。

Nansha Gymnasium, the site of the 2010 Guangzhou Asian Games martial arts competition and Kabaddi competition, has a total construction area of over 30,000 m^2, with a 29 m height, and seating for more than 8,000 including 6,000 fixed seats and 2,000 mobile seats.

Combining the architectural plan configuration with the Taiji (Tai Chi) diagram gives us the opportunity to skillfully use the concept of “martial arts” to organize the project. In the design, nine overlapping surfaces form the shell of the stadium; centered on the hall, they are divided into two groups, north and south, expanding in a radiating spiral. The purpose of this treatment is to divide the single building volume into two, and then re-combine them in a dynamic way, using the composition of the Taiji Yin and Yang diagram as a metaphor of the highest level of Chinese martial arts- “combination of Yin and Yang, man and heaven”.

In addition, we bring in features of the unique regional culture in Guangdong, that is, a marine culture, by drawing on the “conch” shell, which is rich in textural changes, as an inspiration for the design. The architectural form acknowledges the regional culture through the exterior treatment of overlapping nested shells and changing colors on the metal roof.

2010 年亚运柔道 / 摔跤馆（广州大学城华工体育馆）

Judo & Wrestling Gymnasium of 2010 Asian Games (Gymnasium of SCUT, Guangzhou University Town)

项目位置：广东省广州市
设计时间：2008–2009 年
竣工时间：2010 年
建筑规模：12,783 m^2

Location: Guangzhou, Guangdong
Design: 2008–2009
Completion: 2010
GFA: 12,783 m^2

华工体育馆从校园总体规划的角度考虑，将建筑单体和景观统一设计，建筑结合地形，体育馆主体嵌入基地原有土台中，削减了体育馆庞大的体量，巧妙解决了道路与看台之间高差的过渡。体育馆由四个扭壳组合而成的反曲屋面呈现出流畅的造型和蓬勃向上的生命力。在四个支撑构架相交的屋面最高点处，设置了一个配备排风机的拨风构件，既是体育馆内部通风换气的出口，也成为屋面造型的亮点，突出体育馆纪念性形象。代表着华南理工大学老校区重要形象的民族性建筑的柱廊、反宇屋面以及屋角起翘等建筑元素，在新体育馆设计中被现代建筑语言重新演绎，具有异曲同工之妙。

Based on the overall campus planning, Gymnasium of SCUT has unified architectural design and landscape design. The building complies with the terrain, and the main body is embedded into the original hill of the base, which reduces the huge volume of the gymnasium and solves the transition of the height between the road and the stands. The recurved roof of the gymnasium composed of four twisted shells presents a smooth shape and vigorous vitality. At the highest point of the roof where the four supporting frames intersect, an exhaust component equipped with an exhaust fan is set up, which is not only an outlet for ventilation in the stadium, but also a highlight of the roof modeling, highlighting the monumental image of the stadium. Architectural elements such as colonnades, upturned eaves and raised corners of the ethnic buildings that represent the important image of the old campus of SCUT, are reinterpreted by modern architectural language and works similarly in the design of the new gymnasium.

2010 年亚运会游泳馆

Swimming Center of 2010 Asian Games

项目位置：广东省广州市
设计时间：2007 年
竣工时间：2009 年
建筑规模：33,331 m^2

Location：Guangzhou, Guangdong
Design：2007
Completion：2009
GFA：33,331 m^2

广东奥林匹克游泳跳水馆是 2010 年广州亚运会游泳跳水项目的主要比赛场馆之一，在此进行了亚运会游泳、跳水和现代五项游泳比赛及亚残会的游泳比赛。游泳跳水馆用地 9.9hm^2，总建筑面积约 33,300 m^2，总座席数为 4500 座，馆内分别设置了一个 50m×25m×3m 标准比赛池、一个 50m×25m×1.4~1.8m 训练池和一个 25m×25m×5.5m 跳水池。

游泳跳水馆位于广东奥林匹克体育中心基地内，规划位置处于基地的边缘，临近形态张扬的奥林匹克体育场，游泳馆用地东北侧还有同时建设的亚运网球中心，建筑群体空间的不定性十分突出。经过分析与比较，我们采用了最简约的体量处理办法，用完整的矩形来应对局促的基地与复杂的周边建筑环境要求。

本游泳馆在工程造价限额极低的情况下能够做到满足亚运会的使用需求同时兼顾赛后运营实属不易。亚运会举办期间，赛事运行流线清晰，分区合理，给赛事组织带来了相当大的便利性，比赛期间受到亚组委各部门的一致肯定。赛后对外开放使用、日常维护及训练期间管理方便，能耗低。

The Guangdong Olympic Swimming and Diving Stadium was one of the major competition venues of the 2010 Guangzhou Asian Games. The Asian Games contested in swimming, diving, and modern pentathlon, as well as the Paralympics swimming competitions, were held here. The Stadium covers 9.9 hectares, with a total construction area of about 33,300 m^2, and a total seating capacity of 4,500. The stadium includes a standard 50 m × 25 m × 3 m competition pool, 50 m × 25 m × 1.4~1.8 m training pool and a 25 m × 25 m × 5.5 m diving pool.

The Swimming and Diving Stadium is located on the site of the Guangdong Olympic Sports Center, at one edge of the site, near the grand Olympic Stadium. To the northeast is the Asian Games Tennis Center, which was built at the same time. This simultaneous construction schedule meant that there were many uncertainties about the final spatial context during the design phase. Through analysis and comparison, the most straightforward volumetric method-a pure rectangle-was chosen to meet the environmental requirements of a limited site area and complex surrounding structures.

Given the extremely limited project budget, it was not a simple task for this swimming stadium to meet both the requirements of the Asian Games, and also take post-game operations into account. During the period of the Asian Games, competitions went on smoothly with a clear schedule and rational zone division, bringing considerable convenience to the competitions and wining unanimous recognition of all departments of the Asian Games Organizing Committee. The management of post-game public opening up, daily maintenance, and training is convenient, with low energy consumption.

天津科技大学体育馆

The Gymnasium in Tianjin University of Science and Technology

项目位置：天津市
设计时间：2012—2016 年
竣工时间：2017 年
建筑规模：24,352 m^2

Location：Tianjin
Design：2012–2016
Completion：2017
GFA：24,352 m^2

天津科技大学体育馆，是在第十三届全运会的号召下建设于高校中的全新场馆。项目的设计明确了全运会的赛时使用与高校赛后使用并举的目标，使设计既符合全运会比赛的专业性、高标准要求，又满足高校长期使用的多功能、轻维护需求。

设计巧妙地在用地中心多向轴线的交汇处引入圆形开敞空间，成功处理好场地中多个方向的空间和景观轴线关系，灵活地将建筑形式、使用功能以及校园景观地融为一体，给刚硬的体育馆带来一丝圆润柔美的动感气息，使总体布局在理性和谐的前提下不失有机。

体育馆的各场馆单元围绕此圆形开敞空间扇形展开，并顺应校园路网和边界，生成简洁而理性实用的方形体量，构成内圆外方的总体布局形式。其简洁方正的造型传达了高校教育希冀的求真淳朴、含蓄典雅风格。

经过第十三届全运会艺术体操和蹦床项目比赛场地的检验，以及学校近几年的使用，天津科技大学体育馆总体上获得了各方好评，已然成为了天津科大充满活力的校园生活中心。

The Gymnasium in Tianjin University of Science and Technology is a new gym built in the university under the call of the 13th National Games. The design of the project clarifies the goals to guarantee the use of National Games and the daily use of the university after the Game, so that the design meets the professional and high standards of the National Games, and meets the multi-functional and easy-to-maintain needs of the university for long-term use.

The design subtly introduces a circular open space at the intersection of the multi-axis of the center of the site, successfully handles the relationship between the space and the landscape axis in multiple directions in the site, and flexibly integrates the architectural form, the function of use and the landscape of campus. It brings a sleek and dynamic atmosphere to the rigid gymnasium, so that the general layout is organic under the premise of its rationality and harmony.

The units of the gymnasium are fan-shaped around this circular open space, and follow the road and boundary of the campus to generate a simple, rational and practical cube, which constitute the general layout form with a circle inside and squares outside. The simple and square shape of the Tianjin University of Science and Technology conveys the simple and elegant concept that college education aspires to express.

After the test of being the venues of the rhythmic gymnastics and trampoline project in the 13th National Games, as well as the school's use in recent years, the Tianjin University of Science and Technology Gymnasium has won praises from all parties, and has become the vibrant campus center of the school.

河海大学综合体育馆
Hohai University Comprehensive Gymnasiu

项目位置：江苏省南京市
设计时间：2009 年
竣工时间：2012 年
建筑规模：22,100 m^2

Location: Nanjing, Jiangsu
Design: 2009
Completion: 2012
GFA: 22,100 m^2

河海大学综合体育馆位于南京河海大学江宁校区，项目用地面积 13,600 ㎡，总建筑面积 22,100 ㎡，建筑高度 25.9m（单层），建筑层数 1~4 层。

一方面，本项目可建设用地为原规划用于体育馆的一片 1.36 万㎡的场地；另一方面，体育馆的规模随学生人数增长需要扩大至 2997 座，同时一同建设一些文娱设施，包括一个 1100 座剧场、学生活动中心以及学生文化创业用房等，以提供一个能承担大型集会、体育比赛、学生体育锻炼、文娱活动等功能的综合文体设施。显而易见，对于以单层空间为主、体量 2.2 万㎡的综合体育馆，用地捉襟见肘。

根据现场条件，没有将各个场馆分别独立设计，而是引入了综合体的设计概念，将所有内容作为一个整体进行考虑，将可以互补共用的功能区共享使用，将不同的功能单元按照各自的特点相互联系，构成一个有机的整体。

The comprehensive gymnasium of Hohai University is located in Jiangning Campus of Hohai University. The project area is 13,600 m^2, the total building area is 22,100 m^2, the building height is 25.9 m (single storey), and the number of stories is 1~4.

On one hand, the construction land is a 136,000 square meters space originally planned for the gymnasium; on the other hand, the size of the gymnasium needs to be expanded to 2997 with the increase of the number of students; at the same time, some recreational facilities will be built together, including 1,100 seats theaters, student activity centers, and students'cultural and entrepreneurial housing, in order to provide a comprehensive cultural and sports facilities for large-scale gathering, competitions, students'physical exercise and recreational activities. Obviously, for a comprehensive gymnasium with a single-storey space as the main body and a dimension of 22,000 square meters, the land is stretched to the limit.

According to the site conditions, the project does not design each venue independently, but introduces the concept of integrated design, considers all the contents as a whole, complements and shares the use of functional areas, organizes different functional units in accordance with their own characteristics, to form an organic whole.

东莞市长安镇体育公园体育馆

Dongguan Chang'an Gymnasium in Chang'an Sports Park

项目位置：广东省东莞市
设计时间：2011 年
竣工时间：2014 年
建筑规模：22,991 m^2

Location: Dongguan, Guangdong
Design: 2011
Completion: 2014
GFA: 22,991 m^2

长安体育馆作为东莞长安镇标志性建筑，建成后成为东莞长安镇高级别体育比赛、业余体育队训练、群众体育锻炼以及文艺演出的多功能文体活动场所。

项目位于长安体育公园轴线北端，面向莲湖路，通过总图螺旋线的规划，使城市公园景观成为建筑的外部延伸，而建筑成为城市公园景观的焦点。体育馆在有限的用地里，采用多馆合一的集约方式，做到赛时和平时综合利用，使体育馆空间得到最大化利用。造型上通过采用螺旋的平面将各类体育功能空间统一为一个整体，继而生成独特的海螺造型。设计由内而外，旋转放大的功能平面、造型、景墙乃至室外景观，统一考虑浑然一体。比赛厅 88 米跨度的钢结构，是当时国内跨度最大的单层网壳钢结构。

As an iconic building in Chang'an Town, Dongguan, Chang'an Stadium is a high-level multi-functional cultural and sports venues for sports competition, amateur sports team training, exercising and performances.

The project is located at the north end of the axis of Chang'an Sports Park, facing Lianhu Road. Through the planning of the spiral of the general plan, the urban park landscape becomes the external extension of the building, and the building becomes the focus of the urban park landscape. In the limited use of the stadium, the gymnasium adopts a multi-institutional intensive mode to make comprehensive use of competition and spare time to maximize the space of the gymnasium. In the styling, the various sports function spaces are unified into one whole by adopting a spiral plane, and then a unique conch shape is generated. Designed by inside and outside, the function plane, shape, wall and even the outdoor landscape, which are enlarged by rotation and unified into one. The steel structure of the 88-meter span of the competition hall was the single-story reticulated steel structure with the largest span in China at that time.

武汉大学卓尔体育馆

Gymnasium of Wuhan University

项目位置：湖北省武汉市
设计时间：2014—2016 年
竣工时间：2018 年
建筑规模：37,188 m^2

Location：Wuhan, Hubei
Design：2014–2016
Completion：2018
GFA：37,188 m^2

项目位于武汉大学内，由于武汉大学校园风貌的独特性，我们跳脱了将体育馆作为"标志物"的惯性思维，让体育馆以更加谦和的态度与环境对话，作为校园肌理和文脉的延伸。在建筑造型上，我们确定以简洁、内敛的性格为定位，汲取了学校特有的文化特征，采用叠落的坡屋面形式，既与周围历史建筑相映成趣，又在一定程度上削弱了建筑体量。

The project is located in Wuhan University. Due to the uniqueness of the campus style of Wuhan University, we have escaped the inertia of using the stadium as a "marker", allowing the stadium to dialogue with the environment with a more modest attitude as an extension of campus texture and context. In the architectural style, we have determined the orientation of simple and introverted character, which takes the unique cultural characteristics of the school and adopts the form of stacked slope roofs, which not only matches the surrounding historical buildings, but also weakens the building volume to some extent.

柳州市李宁体育馆
Li Ning Stadium in Liuzhou

项目位置：广西壮族自治区柳州市
设计时间：2008 年
竣工时间：2011 年
建筑规模：23,049.15 m^2

Location：Liuzhou, Guangxi
Design：2008
Completion：2011
GFA：23,049.15 m^2

广西柳州素以“中华石都”闻名，所谓“山无石不奇，水无石不清，园无石不秀，室无石不雅”。体育馆的造型理念“以石为形，以石为理，以石为寓”：将主场馆与训练场塑造为两石的形态，通过大小、明暗、材质的对比展现出层次丰富的整体形象；运用现代建筑的材料元素来抽象地表达奇石的条形纹理，使体育馆富有现代气息；奇石形象象征了运动健儿坚韧不拔的拼搏奋斗精神，使体育馆成为一种精神文化的物质载体。

Guangxi Liuzhou is known as the “Chinese stone capital”, where there is an old saying, “Without stones, the mountain would not be magnificent, the water would not be clear, the garden would not be boutique, and the room would not be elegant”. The styling concept of the gymnasium is “the shape of stone, the idea of stone, and the spirit of stone”. The main stadium and the training ground are shaped into two stone shapes, and the overall image is rich in layers through the contrast of size, light and darkness. The material elements of modern architecture are used to abstractly express the strip-shaped texture of the stone, which makes the stadium full of modern atmosphere. The image of the stone symbolizes the hard-working spirit of the athletes and makes the stadium a material carrier of spiritual culture.

常州工学院体育馆

Gymnasium of Changzhou Institute of Technology

项目位置：江苏省常州市
设计时间：2012 年
竣工时间：2016 年
建筑规模：14,131 m^2

Location：Changzhou, Jiangsu
Design：2012
Completion：2016
GFA：14,131 m^2

体育馆设计紧扣高校体育建筑的使用特点，结合了体育竞技、体育教学、体育娱乐、体育锻炼、大型集会活动等的多功能使用特点。大型开敞空间可以灵活分割适应各种活动，在空间上可以灵活多变。

The design closely adheres to the using characteristics of sports buildings in colleges and universities, and combines the multifunctional using characteristics of sports competition, physical education teaching, physical entertainment, physical exercise, large-scale assembly activities and so on. Large open space can be divided to adapt a variety of activities, which can be flexible and changeable in space.

内蒙古行政学院文体活动中心

Sports Center of Inner Mongolia Administration College

项目位置：内蒙古自治区呼和浩特市
设计时间：2013 年
竣工时间：2018 年
建筑规模：12,979 m^2

Location：Hohhot，Inner Mongolia Autonomous Region
Design：2013
Completion：2018
GFA：12,979 m^2

文体活动中心由篮球羽毛球综合馆、网球馆、游泳馆和各种文体活动室组成，是一座综合性多功能的文体活动中心。设计过程中，建筑师将可持续设计的理念贯穿于建筑物的全寿命周期，并且充分考虑了它和周边校园空间、城市肌理的关系。建筑师力求在有限而紧凑的用地内，创造一个内部功能多样紧凑、高效舒适的设施，为学校及周边社区居民的运动健身、文艺活动提供了尽可能多的选择，以实现长久、持续的运营。

The center consists of basketball and badminton hall, tennis hall, swimming hall, and rooms for cultural and sports activities, which is a multi-functional culture and sports center. The center follows sustainable design concept throughout the entire life cycle of the building, with full consideration given to its relationship with surrounding campus space and urban context. Given a limited and compact site, the architect aims to realize an efficient and comfortable establishment with diverse and compact functions, so as to provide more options of sports and fitness, cultural and artistic activities for both the school and nearby residents and achieve long-term and sustainable operation.

华南理工大学松花江路历史建筑保护更新（何镜堂工作室）

Renewal of Historic Buildings in South China University of Technology (He Jingtang Studio)

项目位置：广东省广州市
设计时间：2004 年
竣工时间：2011 年
建筑规模：2,600 m^2

Location：Guangzhou, Guangdong
Design：2004
Completion：2011
GFA：2,600 m^2

07 其他建筑

OTHER ARCHITECTURE

华南理工大学松花江路历史建筑保护更新（何镜堂工作室）	Renewal of Historic Buildings in South China University of Technology (He Jingtang Studio)
海棠湾 9 号度假酒店	No.9 Haitang Bay Gloria Resort Sanya
南湖山庄 C 区别墅与 D 区高层住宅	Nanhu Villa District C Villa and D District High-rise Residence
横琴新家园首期住宅项目	Hengqin New Home First Phase Housing Project
鹤山峻廷湾一期	Heshan Junting Bay Phase I
广州大学城 DS1601、DS1604 地块建筑设计	DS1601 and DS1604 Blocks in Guangzhou Higher Education Mega Center
佛山南海 Aloft 酒店	Aloft Nanhai Foshan
七仙岭温泉度假村	Qixianling Hot Spring Resort
海丰县第二人民医院	Haifeng Second People's Hospital
临夏州人民医院	Linxia State People's Hospital
省残疾人康复基地首期工程	Provincial Rehabilitation Base for the Disabled Phase Ⅰ
广州火车东站天幕广场	Guangzhou East Railway Station Tianmu Square
增城区人民法院	Zengcheng People's Court
广州市第五资源热力电厂二期工程及配套设施	Guangzhou Fifth Resource Thermal Power Plant Phase Ⅱ Project and Supporting Facilities
横琴新区区域供冷站和冷冻水管网工程第三供冷站	District Cooling Station & Cooling Pipe of Hengqin New Area NO.3 Cooling Station
韶关武江乡村振兴传习所	Rural Vitalization Training Center of Wujiang, Shaoguan

成都火车站扩能改造

Expansion of Chengdu Railway Station

项目位置：四川省成都市
设计时间：2010 年
竣工时间：2024 年（预计）
建筑规模：245,067 m^2

Location: Chengdou, Sichuan
Design: 2010
Completion: 2024 (expected)
GFA: 245,067 m^2

成都是四川省会，成都站除了要满足它作为交通枢纽的功能，更应该成为传播四川地域文化的门户。成都站位于城市轴线北端，包含城市文化特色的主体建筑以一种典雅的现代“城市门楼”的形象出现，既是对城市特有文化气质的表达，也大气呼应了城市的轴线格局。

本案以蜀竹和芙蓉花为灵感，构成形式和结构的基本单元，并加以组合，构建出具有文化意蕴并符合结构逻辑的整体形象。立面形象中的伞形单元抽象地展现出成都市花“芙蓉”的集合意向，成都站犹如城市轴线上一朵盛开的芙蓉花，丰姿绰约。

Chengdu is the capital of Sichuan Province. Besides being an important transportation hub, Chengdu Railway Station should also be a gateway to spread Sichuan regional culture. Chengdu Railway Station is located at the northern end of the urban axis, and the main building, featuring the city's cultural characteristics, appears as an image of an elegant "city gatehouse", which not only expresses the unique cultural temperament of the city, but also echoes the outline of the city. This project draws the inspiration from Sichuan bamboo and hibiscus flowers, turning them into a basic unit of form and structure, and combining them in various ways to construct an image of cultural meaning and structural logic. The "umbrella" within the elevation image abstractly shows the collective intent of the hibiscus flower (the official flower of Chengdu) thus turning Chengdu Railway Station into a blooming hibiscus flower on the city axis.

前海公共空间设计
Qianhai Public Spaces Design

项目位置：广东省深圳市
合作单位：北京土人城市规划设计股份有限公司、
东南大学建筑设计研究院有限公司
设计时间：2019 年
竣工时间：2022 年（预计）
建筑规模：8,200 m^2

Location：Shenzhen, Guangdong
Partner：Turenscape, Architects & Engineers Co, Ltd. of Southeast University
Design：2019
Completion：2022 (expected)
GFA：8,200 m^2

项目包含四座融于城市、连接自然、空间丰富、展现前海特色风貌的跨街公园与天桥。为符合前海宜居乐业、创新智慧、高效便捷、绿色生态的城市定位，应对用地周边高效紧凑、复杂多样，公共交通密集，使用人群目的多样的复杂城市环境，设计以打造“引领创新、体现先进”示范性标杆为基调，提出以下三大设计策略：

协调性策略——融于城市：方案尊重现状城市特色，以“隐”为基本定位，谦逊地融于城市，顺应现有景观要素，融入绿地体系。

整体性策略——和而不同：通过共性特点与个性特点，打造有机协调的形态与景观空间，根据对所在片区人群目的多样化的研究制定个性化空间。

先进性策略——引领创新：采用先进结构、轻质种植技术，以及人性化、智能化技术与设备等先进技术与理念着力打造公共空间的创新体验与特色。

The project contains four street-cross parks and flyovers that integrate the city, connect nature and display the distinctive features of Qianhai. The orientation of Qianhai is livable and work in peace, innovation and intelligence, efficient and convenient, green and ecological. The surrounding land is efficient, compact, complex and diverse. It also has public transportation and users of various purpose. Facing to the complex urban environment, the project is based on the demonstration of “leading innovation, reflecting advance”, putting forward the following three design strategy.

Coordination strategy-integration into the city. The plan respects the current urban characteristics, takes “hidden” as the basic positioning, and humbly integrates into the city, conforming to the existing landscape elements and integrating into the green space system.

Holistic strategy-harmony without uniformity. The plan creates organically coordinated forms and landscape space through common characteristics and individual characteristics, and develops personalized space according to the diversified research on the population and eye in the area.

Advanced strategy——leading innovation. The plan uses advanced structure, lightweight planting technology, humanized and intelligent technology and equipment and other advanced technologies and concepts to create innovative experience and characteristics of the public space.

对应广州国际化大都市的定位，加上桥梁选址位置的特殊与重要性，很明显这将不是一座普通的桥梁，因此设计充分顺应现有城市肌理，采用曲线形设计融入广州中轴花瓶曲线的总体设计，呼应周边城市景观。主体花舟景观抽取岭南花舟的设计形象作为图底，结合台地花卉的布置，运用太阳能光伏、清凉喷雾、智慧桥梁技术等现代创新科技，打造独具广州特色，适应岭南气候，通行体验舒适的岭南花桥。桥梁建成后将成为世界上跨度最大、宽度最大的斜拱曲梁跨江人行景观桥。

Corresponding to the positioning of Guangzhou as an international metropolis and the important location of the bridge, it will not be an ordinary bridge. The design fully conforms to the existing urban texture, and adopts the curved design to integrate into the overall design of the central axis vase curve in Guangzhou, echoing the surrounding urban landscape. The design image of Lingnan flower boat is selected as the main part of the landscape. Combined with the arrangement of flowers on the platform, the bridge uses modern innovative technologies such as solar photovoltaic, cooling spray and intelligent technology. It will be created as a Lingnan flower bridge of unique Guangzhou characteristic, with the adaptation to the local climate and comfortable pedestrian experience. It will be the world's largest span and width of the cable-stayed arch pedestrian landscape bridge across the river.

广州海心桥
Guangzhou Haixin Bridge

项目位置：广东省广州市
合作单位：广州市市政工程设计研究总院有限公司
设计时间：2019 年
竣工时间：2021 年
建筑规模：7,000 m^2，主跨 196 m，宽度 15 m

Location：Guangzhou, Guangdong
Partner：Guangzhou Municipal Engineering Design & Research Institute Co,Ltd.
Design：2019
Completion：2021
GFA：7,000 m^2，Main span：196m，Width：15m

广州白云站位于广州白云区的南侧，规划定位为广州枢纽主要客站之一，在铁路客运系统中占有重要地位。

白云站站房设计秉承TOD开发理念，努力打造一个畅通融合、绿色温馨、经济艺术、智能便捷的具有岭南特色的全新铁路枢纽客站。方案以展现广州城市名片与绿色生态的设计理念为核心，构筑标志性城市形象。实体为山：以舒展起伏的建筑形态与不同标高层层叠级的绿色平台隐喻白云山。虚体为水：以建筑内流动的光线与交通人流寓意珠江水静静流淌生生不息。建筑中的"光谷"成为最具特色的标志性空间，片片墙体弧形延展至屋面，营造出木棉花开的美丽形态，同时也可为绿化阶梯坡道遮蔽风雨，为候车室隔绝东西向日晒。

设计中提出"可伸缩的车站、会呼吸的广场"概念。整体项目采用"方—圆—方"的图底关系布局，其中外方为城，内方为站，方与圆之间是两个呼吸广场。呼吸广场平时可作为舒适宜人的休闲景观广场，春运等特殊状况时成为容纳大量旅客的候车空间。

Guangzhou Baiyun Railway Station is located on the south side of Guangzhou Baiyun district and is planned to be one of the main stations of Guangzhou hub. Therefore, it occupies an important position in the railway passenger transport system.

The station house design of Baiyun station adopts the concept of TOD development and strives to create a new railway terminal with Lingnan characteristics, which is smooth and integrated, green and warm, economic and artistic, intelligent and convenient.

The scheme takes the design concept of showing Guangzhou city name card and green ecology as the core and constructs the iconic city image. The entity is a mountain, and the white cloud mountain is depicted by the undulating architectural form and the green platform with different levels of high-rise buildings. The virtual body is water -the light and traffic flow in the building compare the image of the Zhu River flowing quietly and continuously. The "valley of light" in the building has become the most distinctive landmark space. The walls extend all the way to the roof, creating the beautiful form of kapok. At the same time, this can also shelter the rain and wind for the green ladder ramp and insulate the waiting room from the problem of sunlight in the east direction.

The concept of "retractable station and breathing square" was proposed in the design. The overall project adopts the "square-round-square" bottom relation layout, in which the outside is the city, the inside is the station, and there are two breathing squares between the square and the circle. Breathing square can be used as a comfortable leisure landscape square. At the same time, in the special condition of Spring Festival travel, it can be a waiting space for a large number of passengers.

白云站

白云站

广深铁路石龙站迁建工程

Relocation of Shilong Station of Guangshen Railway

项目位置：广东省东莞市
设计时间：2009 年
竣工时间：2014 年
建筑规模：20,664 m^2

Location：Dongguan, Guangdong
Design：2009
Completion：2014
GFA：20,664 m^2

建筑造型提取岭南民俗“赛龙舟”的文化意向，抽象出深入民心的岭南龙舟形象与岭南民居造型元素“龙船脊”造型特征为主要素材，形成“龙舟竞渡”的建筑意向，象征石龙与新莞人锐意进取、奋勇拼搏的精神。

建筑造型简洁、舒展、大方，充分体现地域及人文特征的同时，也体现了铁路交通建筑实用、高效的特点。

The architectural style extracts the cultural intention of the Lingnan folk custom “dragon boat”, abstracts the image of Lingnan Dragon Boat and the character of the “dragon boat ridge” of Lingnan residential building as the main material, forming the architectural intention of “dragon boat race”, symbolizing Shilong residents and migrants of Dongguan are determined to forge ahead and work hard.

The architectural style is simple, stretched and generous, fully embodying the regional and human characteristics, and also reflects the practical and efficient characteristics of railway traffic construction.

江西省抚州站

Fuzhou Station, Jiangxi Province

项目位置：江西省抚州市
设计时间：2009 年
竣工时间：2013 年
建筑规模：11,998 m^2

Location: Fuzhou, Jiangxi
Design: 2009
Completion: 2013
GFA: 11,998 m^2

抚州站以古代书简、卷轴为主要素材，结合当地深厚的临川文化底蕴，设计提炼出“轴卷书开、文迎四方”作为建筑造型的文化意象。建筑简洁大气，四个柱子是从中华民族优秀文化的一位戏剧巨人汤显祖的代表作“临川四梦”中抽象出来的，张开的书简像是向人们诉说着千年历史的临川文化，充分体现了当地的人文特征。

Fuzhou Station uses ancient books and scrolls as the main material, combined with the local Linchuan cultural heritage, to design and extract the cultural image of architectural style. The building is concise and the four pillars are abstracted from the representative of Tang Xianzu, “The Four Dreams of Linchuan”, a drama master of the Chinese nation’s outstanding culture. The open book is like a Linchuan culture that tells people about the millennium history. It reflects the local human characteristics.

人民路高架改造及沿线品质提升概念策划

Landscape Innovation of Renmin Overpass Road

项目位置：广东省广州市
合作单位：广州市城市规划勘测设计研究院、广州市市政工程设计研究总院有限公司
设计时间：2019 年
建筑规模：50,000 m^2

Location: Guangzhou, Guangdong
Partner: Guangzhou Urban Planning & Design Survey Research Institute
Guangzhou Municipal Engineering Design & Research Institute Co, Ltd.
Design: 2019
GFA: 50,000 m^2

人民路高架桥地处广州老城核心位置，是中国大陆第一座高架桥，是“广州奔向现代化”的象征，承载了城市变迁。设计师将人民路高架 4.7km 长度根据周边城市区域特点分为四段，各段具备自己独具特色的设计主题，同时各段之间相辅相成，共同构成具有整体性的高线公园。

The Renmin Overpass Road is located in the center of Guangzhou. It is the first overpass road in China. It is a symbol of “Guangzhou's rush to modernization” and bears the city's changes. The Renmin Road elevated 4.7km length is divided into four sections according to the characteristics of the surrounding urban areas. Each section has its own unique design theme. At the same time, each section complements each other to form a holistic high-line park.

广州大桥改扩建景观提升设计

Expansion and Landscape Innovation of Guangzhou Bridge

项目位置：广东省广州市

合作单位：林同棪国际工程咨询（中国）有限公司

设计时间：2017 年

建筑规模：17,000 m^2

Location: Guangzhou, Guangdong

Partner: T. Y. Lin International Engineering Consulting（China）Co, Ltd.

Design: 2017

GFA: 17,000 m^2

广州大桥位于珠江之上，是广州市区主干道路的组成部分。广州大桥于 1985 年 5 月建成通车，2017 年 9 月开始扩建改造工程。设计中选方案以海上丝绸之路为设计理念，结合岭南地区的气候文化特色，为市民提供观赏与通行皆宜的桥梁。

The Guangzhou Bridge is located above the Zhujiang River and is an integral part of the main roads in the urban area of Guangzhou. The Guangzhou Bridge was opened to public in May 1985, and the expansion and renovation project began in September 2017. The concept of the selected design scheme is based on the Maritime Silk Road, combined with the climate and cultural characteristics of the Lingnan region, to provide citizens with a bridge suitable for viewing and passing.

深中跨海通道西人工岛建筑景观工程
West Artificial Island of Shenzhen-Zhongshan Cross-Sea Link

项目位置：珠江出海口
设计时间：2016—2022 年
竣工时间：2024 年（预计）
建筑规模：16,574 m²

Location：Estuary of Zhujiang River
Design：2016–2022
Completion：2024 (expected)
GFA：16,574 m²

该建筑设计有别于常规的公共建筑设计，而是一个以满足桥梁、人工岛与隧道的多方面要求为前提，同时实现景观美学与使用体验的重大交通工程。复杂的外界需求给建筑设计带来了包括结构、给排水、通风、消防、电气等诸多限制。我们通过对超出常规建筑领域的再学习，最终在深中通道西人工岛上实现了技术与美学的融合。我们将在建筑中增加了一个两层高的清水混凝土基座，基座上部为对外开放的餐饮区、展览区以及对内的办公管理区，基座下部主要为设备用房、仓库与消防救援站。

Different from conventional design for public buildings, architectural design of this project has to firstly respond to the various requirements of bridge, artificial island and tunnel. As a key transportation facility, the project needs to realize goals in both landscape aesthetics and user experience. Besides, the complex external demands bring various restraints over the architectural design in terms of structure, water supply and drainage, ventilation, fire, electric, etc. Through the study of buildings exceeding code limits, the design team eventually realized the integration of technology and aesthetics on the project. A two-story fair-face concrete base is added to the building. The upper part of the base provides areas for catering and exhibition open to the public, and internal office management functions, while the lower part is mainly used as mechanical rooms, warehouses and fire staging stations.

成 都

广州铁路枢纽新建广州白云站

Guangzhou Baiyun Railway Station

项目位置：广东省广州市
设计时间：2018
竣工时间：2022 年（预计）
建筑规模：500,000 m^2

Location：Guangzhou, Guangdong
Design：2018
Completion：2022 (expected)
GFA：500,000 m^2

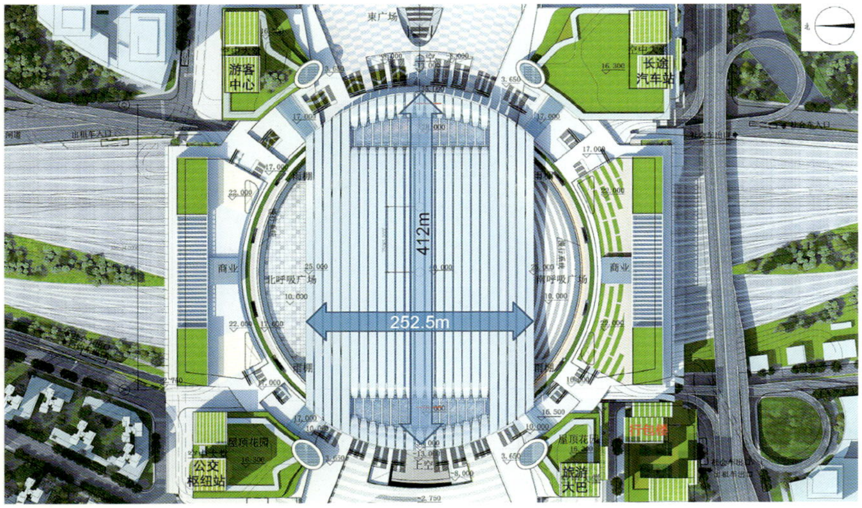

06 交通建筑
TRANSPORT ARCHITECTURE

广州铁路枢纽新建广州白云站 Guangzhou Baiyun Railway Station
广州海心桥 Guangzhou Haixin Bridge
前海公共空间设计 Qianhai Public Spaces Design
成都火车站扩能改造 Expansion of Chengdu Railway Station
深中跨海通道西人工岛建筑景观工程 West Artificial Island of Shenzhen-Zhongshan Cross-Sea Link
广州大桥改扩建景观提升设计 Expansion and Landscape Enhancement Design of Guangzhou Bridge
人民路高架改造及沿线品质提升概念策划 Landscape Revolution of Renmin Overpass Road
江西省抚州站 Fuzhou Station, Jiangxi Province
广深铁路石龙站迁建工程 Relocation of Shilong Station of Guangshen Railway

华南理工大学松花江路 14、18、20、30—37 号，原为 20 世纪中山大学教授的居住区，现已被列为历史保护建筑。在该居住区内，北面一列是 5 栋 1930 年代建成的坡屋顶别墅，南面一列是 4 栋 1960 年代建成的独立式住宅。由于年久失修，这些建筑均已破败，更有部分已成危房。面对这些在城市高速发展中被遗忘的历史建筑，我们从 2004 年起至今，经过三次渐进式的更新改造，采用园林化、机能更新、社区融入、生态节能等设计策略，将这组破败的历史建筑逐渐转变为一个独具岭南地域特色的、充满朝气活力的建筑师工作室，为校园创造了一个富有文化气息的创新基地，为社区创造了一个自然、人性化的公共活动场所。

Buildings at No. 14, 18, 20, 30-37, Songhuajiang Road on the campus of South China University of Technology (SCUT) were once the residences for the professors who working with the former Sun Yat-Sen University in the 20th century and now are listed as the historic buildings. In this residential area, the north part is occupied by five slope-roofed houses built in 1930s, while four detached residential area built in 1960s stand in the south. These buildings have fallen into disrepair, and some have been judged as dilapidated house. From 2004 until now, three rounds of renewal and renovations have been conducted step by step, and these run-down historic buildings which forgotten in the rapidly growing city have been transformed into a vibrant architect studio with unique Lingnan features. The strategies include the gardening, function renewal, community integration and ecological energy-saving. The renewal practices not only create an innovation base with cultural atmosphere on SCUT campus, but also offer a natural and human-oriented public activity place for the community.

海棠湾 9 号度假酒店
No.9 of Haitang Bay Gloria Resort Sanya

项目位置：海南省三亚市
设计时间：2009 年
竣工时间：2012 年
建筑规模：87,000 m^2

Location：Sanya, Hainan
Design：2009
Completion：2012
GFA：87,000 m^2

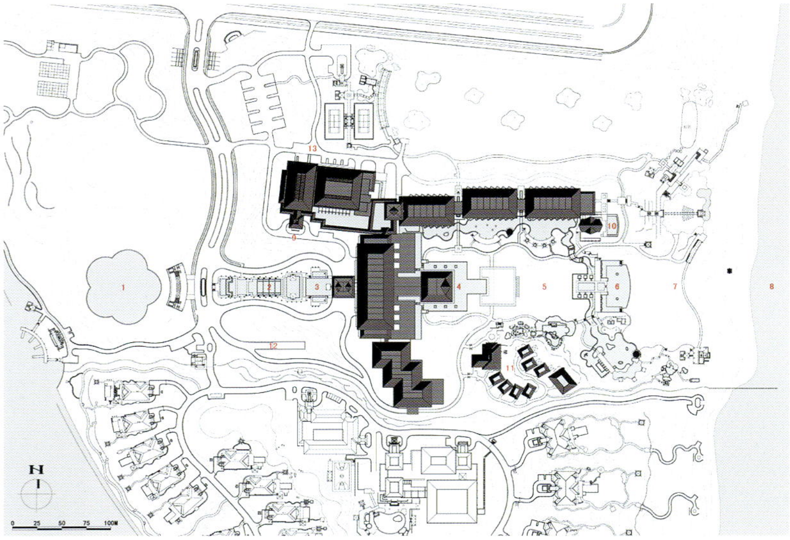

海棠湾 9 号度假酒店地处中国南部的国家海岸，是海棠湾上唯一一座以岭南中式建筑风格为主的度假酒店。它向国际游客展现了中国建筑文化神韵和琼州地域文化风情，向世界诠释中国旅游度假建筑发展的新方向。

海棠湾 9 号度假酒店与阳光、银滩、碧海、青山等自然元素共同勾画出一幅“海天山宇”的水墨山水画。方案分别从造型、构造、材料三方面传承中国建筑文化的艺术特征，采用现代建筑语汇营造富有传统文化和地域风情的建筑空间。

No.9 Haitang Bay Gloria Resort Sanya, located along the national coast of southern China, is the only resort hotel dominated by the Lingnan architectural style in Haitang Bay. It shows the charm of Chinese architecture and Hainan regional culture and customs to international tourists. It is a fine example of new development trends for China's tourist resorts.

No.9 Haitang Bay Gloria Resort Sanya combines natural elements such as sunshine, silver sand beaches, blue seas and green hills to collectively draw an ink landscape painting of the sea, sky, mountains and houses. The design uses artistic features of Chinese architectural culture in the three aspects of modeling, structure and material, and uses modern architectural vocabulary to create an architectural space rich in traditional culture and regional customs.

南湖山庄 C 区别墅与 D 区高层

Nanhu Villa District C Villa and D District High-rise Residence

项目位置：广东省广州市
设计时间：2009 年
竣工时间：2014 年
建筑规模：145,164 m^2

Location: Guangzhou, Guangdong
Design: 2009
Completion: 2014
GFA: 145,164m^2

南湖山庄位于广州市白云区。项目基地为山地地形，场地整体东高西低，东侧山体呈月牙形态势环抱场地。项目共有 A、B、C、D、E 等若干组团，本次设计范围包括 C 区（别墅区）与 D 区（高层住宅区），其中 D 区四栋高层建筑，呈圆弧形排列于半山之上，成为项目中最瞩目的标志性元素。

Nanhu Villa is located in Baiyun District, Guangzhou. The project base is mountainous terrain, the whole site is high in the east and low in the west, and the site is surrounded by the crescent shape mountain in the east. The project has a number of groups including A, B, C, D, E, etc. The design scope includes area C (villa area) and area D (high-rise residential area), of which four high-rise buildings in area D are arranged in a circular arc shape. Above the mountain, it became the most significant element in the project.

横琴新家园首期住宅项目

Hengqin New Home First Phase Housing Project

项目位置：广东省珠海市
设计时间：2010 年
竣工时间：2012 年
建筑规模：143,739m²

Location: Zhuhai, Guangdong
Design: 2010
Completion: 2012
GFA: 143,739m²

横琴新家园是国家开发横琴新区的一项政府安居工程。首期住宅项目主要安置横琴新村、旧村两个自然村的原住民，首期工程项目用地位于新家园整体规划用地的东北面，位于横琴环岛东路以西（澳门大学西侧）、粗沙环以北。工程总建筑面积 143,739.05 m²，由 9 栋 21~30 层高层住宅楼组成，住宅总户数 1077 户。

小区设计强调以人为本，不但总体规划上考虑了回迁村民的交流空间和邻里关系，同时考虑了户型的舒适性与合理性结合，为横琴地区创造出第一个富有岭南特色及海岛风情的村民回迁小区。

Hengqin New Home is a government housing project in Hengqin new area developed by the state. The first phase of the housing project mainly houses the residents of Hengqin New Village and Old Village. The first phase of the project is located in the northeast of the overall planning land of the New Home, west of Hengqin Huandao East Road (west of the University of Macau) and north of Cushahuan. The total construction area of the project is 143,739.05 square meters, consisting of 9 high-rise residential buildings with 21~30 stories , with a total number of 1077 households.

The design of the residential area emphasizes people-oriented, not only considering the communicate space and neighborhood relationship of the returned villagers in the overall planning, but also considering the combination of comfort and rationality of housing type, creating the first residential area full of Lingnan characteristics and island customs for the villagers in Hengqin.

横琴新家园

鹤山峻廷湾一期
Heshan Junting Bay Phase I

项目位置：广东省江门市
设计时间：2010 年
竣工时间：2014 年
建筑规模：89,100 m^2

Location: Jiangmen, Guangdong
Design: 2010
Completion: 2014
GFA: 89,100 m^2

项目设计运用现代建筑风格与形式体现一定的科技人文色彩，演绎现代、时尚、简约、大气的形象，以迎合目标客户群的喜好，体现现代城市文明规划意境。

规划设计上利用地块地下水资源丰富的特点营造中央人工湖，创造内部景观以提升建筑素质。根据地形特点，沿东面河道与南侧规划道路以布置较高密度的排屋为主；而沿东北鱼塘与西北田园，则布置风车形式的多层住宅；在地块西南侧靠近规划道路，与二期发展用地交接处，结合小区出入口布置综合楼；而在小区内部，环境稳定、安静，结合人工湖景，布置面积较大的户型。

The project design uses modern architectural style and form to reflect certain technological humanistic elements, deducing the image of modernity, fashion, simplicity and dignification, so as to meet the preferences of target customers and reflect the artistic conception of modern urban civilization planning.

In terms of planning and design, the characteristics of abundant ground water resources are utilized to create a central artificial lake and to create internal landscape to improve the architecture quality. According to the topographic characteristics, the layout of a relatively high density of houses is arranged along the east river and the south side of the planning road; and the layout of the multi-story house with windmill form is arranged along the northeast fishpond and northwest countryside; in the southwest side of the plot, which is close to the planning road and the junction with the second-phase development land, the comprehensive building is arranged in combination with the gateway of the community; and in the residential area with stable and quiet surroundings, the apartment which has large area is arranged combined with artificial lake scenery.

广州大学城 DS1601、DS1604 地块建筑设计

DS1601 and DS1604 Blocks in Guangzhou Higher Education Mega Center

项目位置：广东省广州市
设计时间：2009 —2013 年
竣工时间：2014 年
建筑规模：127,000 m^2

Location：Guangzhou, Guangdong
Design：2009–2013
Completion：2014
GFA：127,000 m^2

项目中四个并置的地块沿东南方向依次排开，形成总长度接近 700m 的带状用地。设计概念为“千帆并举”，以建筑群雕的方式回应一线临江的场地主题，在规划结构上立足于大学城中轴线界面的城市设计的大格局概念，既强调大学城主轴线的纵向空间序列营造，又为轴线对侧用地的进一步开发提供多样的互动可能。规划上是以地块为单位划分组团，再通过组团的空间沟通与交通联系进一步建立居住社区。各地块的建筑以地形、朝向和景观为依据布局，实现土地开发效益的最大化。各建筑之间是富于变化的自由式布局，每一地块具有主题明确的区域中心，各地块又形成相互贯通的景观链条，力图实现住宅小区空间与城市空间的延续性与尺度转换。

The four plots of the project are arranged in turn along the southeast direction to form a belt-shaped land with a total length of nearly 700 meters. The design concept is “several thousands of sails”, responding to the theme of facing the river in the form of architectural group carvings. The concept of large-scale urban design is based on the central axis interface of the Higher Education Mega Center in the planning structure, which emphasizes the vertical space sequence construction of the main axis. It also provides a variety of interaction possibilities for the further development of the axis-to-side land. The planning is divided into groups by plots. The residential community is established through the space communication and transportation links of the group. The buildings in each block are based on the topography, orientation and landscape, and maximize the benefits of land development. The buildings are rich in free-form layouts. Each block has a regional center with a clear theme, and forms a landscape chain that intersects each other, trying to realize the continuity and scale conversion of the residential community space and urban space.

佛山南海 Aloft 酒店

Aloft Nanhai Foshan

项目位置：广东省佛山市
设计时间：2007 年
竣工时间：2010 年
建筑规模：27,096 m^2

Location：Foshan, Guangdong
Design：2007
Completion：2010
GFA：27,096 m^2

建筑体量方正，采用两种表皮的组合，局部穿插玻璃幕墙，增强虚实变化，整体大气。山墙和主入口布置巨大的酒店标识，结合大型屋顶及入口飘蓬，墙体加以适当的颜色穿插，增强建筑的吸引力。深凹的窗户和竖向构件遮挡阳光直射，塑造出简洁的网格结构，彰显了特有的建筑性格及地域特点。

The volume is regular and graceful, combining two skins and partially interspersed with a glass curtain wall to enhance the changes. The gables and the main entrance are furnished with a huge hotel logo, combined with a large roof and entrance flutter, and the walls are interspersed with appropriate colors to enhance the attraction of the building. The deep concave windows and vertical components block direct sunlight, creating a simple grid structure that highlights the unique architectural character and regional features.

七仙岭温泉度假村

Qixianling Hot Spring Resort

项目位置：海南省保亭县
设计时间：2009 年
竣工时间：2013 年
建筑规模：35,177 m^2

Location: Baoting, Hainan
Design: 2009
Completion: 2013
GFA: 35,177 m^2

设计坚持以人为本，强调环境作主角、建筑作配角，注重生态环境，坚持可持续发展的规划原则。总体规划由一条贯穿南北的小溪流自然分为东西两部分，以主楼为中心，围绕南北向的中轴在东西两侧分别布置别墅式客房楼和集中式客房楼。用地中部为大面积的室外休闲活动空间，集健身游泳区、山泉戏水区和温泉泡池区于一体。

The project design adheres to the people-oriented principle, emphasizes environment as the leading role and the building as a supporting role, pays attention to the ecological environment, and adheres to the planning principle of sustainable development. The master plan is naturally divided into two parts, by a north-south stream, and the main building as the center, around the central axis, villa-style rooms are arranged on the east side and centralized rooms on the west side. The middle part of the land is a large area of outdoor leisure activity space, which integrates fitness swimming area, mountain spring water area and hot spring pool area.

海丰县第二人民医院

Haifeng Second People's Hospital

项目位置：广东省海丰县
设计时间：2017 年
竣工时间：2019 年
建筑规模：36,970 m²

Location：Haifeng, Guangdong
Design：2017
Completion：2019
GFA：36,970 m²

本项目为改造升级工程，在原梅陇镇中心卫生院基础上，坚持可持续发展的设计理念、以人为本的设计原则，着力于对原有医疗资源的整合以及营造优质的医疗环境。拆除年代较为久远的危楼，保留较新的门诊楼与住院楼，新建建筑以 L 形插入到保留建筑中，既整合了原有建筑，又在总体场地上自然划分出入口前广场与住院庭院两大块室外空间，形成中轴统领的空间结构，新旧建筑相互融合，和谐共生。

This project is a transformation and upgrading project. Based on the original Meilong Town Health Center, we adhere to the sustainable development design concept and people-oriented design principles, focusing on the integration of original medical resources and creating a high quality medical environment. Demolishing the dangerous buildings of a relatively old age, retaining the new clinic building and inpatient building, the new building is inserted into the reserved building in a L shape. This way not only integrates the original building and naturally divides the front entrance square and the hospitalized courtyard on the overall site, but also forming the spatial structure of the central axis, making new and old buildings blend together, harmoniously.

临夏州人民医院

Linxia State People's Hospital

项目位置：甘肃省临夏回族自治州
设计时间：2013 年
竣工时间：2018 年
建筑规模：85,118 m^2

Location：Linxia, Gansu
Design：2013
Completion：2018
GFA：85,118 m^2

临夏州人民医院是一所集医疗、预防、保健为一体的 1200 床现代化三级综合医院。在用地紧张、投资有限的条件下，设计秉持高效、绿色、人文三大理念。医疗综合楼通过贯穿整个建筑的医院街来便捷联系门诊、医技和住院三大部分。在各医疗单元之间则引入形式多样的庭院，为候诊廊、诊室、医护工作区提供自然通风和采光。门诊大厅设计为一个四层通高的阳光大厅，与医院街相对独立又紧密联系，成为组织患者就诊的重要空间。

Linxia State People's Hospital is a 1200-bed modernized general hospital, integrating medical treatment, prevention and health care. Under the conditions of limited land and investment, the design adheres to the three concepts of high efficiency, greening and humanity. The medical complex provides easy access to clinics, medical skills and inpatient department through hospital streets throughout the building. A variety of courtyards are introduced among the medical units to provide natural ventilation and daylighting for the waiting rooms, clinics and medical work areas. The outpatient hall is designed as a four-story sunshine hall, which is relatively independent and closely connected with the hospital street, and becomes an important space for organizing the patients.

省残疾人康复基地首期工程

Provincial Rehabilitation Base for the Disabled Phase I

项目位置：广东省广州市
设计时间：2013 年
竣工时间：2017 年
建筑规模：90,000 m^2

Location: Guangzhou, Guangdong
Design: 2013
Completion: 2017
GFA: 90,000 m^2

该建筑构想以“希望的梯田”为主题概念，“梯田”是山区劳动人民在艰苦条件下高超智慧与顽强精神的产物，本方案旨在借“梯田”精神激励残疾人克服自身缺陷的困扰，凭借自身顽强意志，在社会多方的关注下积极治疗，最终康复并重新融入社会之中，而康复基地本身也希望与残疾朋友一起在人生的阶梯中收获硕果。与此同时，“梯田”的形态结构也与生态化的主体结合，适时引入外界光源使室内空间映出绿色气息，营造出恬静的疗养空间。

The concept of the building is based on the concept of “desirable terrace”. The “terrace” is the result of the high level of wisdom and tenacity of the working people in mountainous areas under difficult conditions. The program aims to encourage the disabled to overcome their own shortcomings by the spirit of “terraced”, keep their stubborn will, seek treatment actively under the concern of the society, and finally recover and reintegrate into society. As a rehabilitation base, it also hopes to harvest fruitful results in the ladder of life together with the disabled. At the same time, the morphological structure of the “terrace” is also combined the main body of the ecologicalization, the timely introduction of external light sources make the indoor space reflect the green atmosphere, creating a quiet healing space.

广州火车东站天幕广场

Guangzhou East Railway Station Tianmu Square

项目位置：广东省广州市
设计时间：2009—2016 年
竣工时间：2016 年
建筑规模：8,269 m^2

Location：Guangzhou, Guangdong
Design：2009–2016
Completion：2016
GFA：8,269 m^2

项目设计始于 2009 年，以广州市新中轴线整体城市设计为背景，建立一个人性化、高效的城市门户区。火车东站门户节点的改造设计不仅在使用功能上进行疏导和优化，还针对乘客抵离体验、建筑美感与低能耗使用方面进行深入的思考与实践，从城市设计与都市景观的角度优化城市门户、枢纽公共场所的空间品质。2016 年的后续设计加入了相对人性尺度的建筑元素，越来越多的公共活动开始出现，场所的公共性逐步恢复。

Standing on the central axis of Guangzhou, the project is decorated with Guangzhou's cityscape as its background. The aim of the project, started from 2009, is to meet the functional requirements of the railway station, and to upgrade this area into a more human-oriented and efficient gateway. There must be new and multi-layered roles, such as a node for tourism, entertainment and commerce. The specific design not only reorganizes and optimizes the programs, but also takes furthers steps in the experience, sustainability and architectural aesthetic, to improve the quality of environment with urban strategies. In 2016, there is a new round of renovation to be launched. To shrink down the enormous scale, more small elements are added. As more and more public activities reappears, the public space is now in its recovery.

增城区人民法院

Zengcheng People's Court

项目位置：广东省广州市
设计时间：2013 年
竣工时间：2019 年
建筑规模：38,641 m^2

Location: Guangzhou, Guangdong
Design: 2013
Completion: 2019
GFA: 38,641 m^2

增城区人民法院位于广州增城区荔城大道，用地东南高、西北低。设计顺应地形高差，通过大基座、大法庭和建筑主体三部分相结合，以“鼎”为原型，对称布置，厚基座与现代倾斜立面塑造出宏大、端庄、威严的法院建筑形象，寓意“一言九鼎”，彰显依法治国的法律权威。

Zengcheng People's Court is located in Licheng Avenue, Zengcheng District, Guangzhou. The terrain is high in the southeast and low in the northwest. The design conforms to the elevation difference of the terrain, and combines three parts, namely the large base, the large court and the main building. Taking the tripod as the prototype, the design adopts symmetrical layout, the thick base and the modern sloping facade shape the grand, and dignified building image of the court, which means “one word is as heavy as nine tripods” to highlight the legal authority of the rule of law.

广州市第五资源热力电厂二期工程及配套设施

Guangzhou Fifth Resource Thermal Power Plant Phase II Project and Supporting Facilities

项目位置：广东省广州市
设计时间：2019 年
竣工时间：2021 年
建筑规模：89,969 m^2

Location: Guangzhou, Guangdong
Design: 2019
Completion: 2021
GFA: 89,969 m^2

广州市花都区第五资源热力电厂二期工程设计以“高山流水，锦绣花都”为主要创作构思，整体厂房采用“去工业化”设计，艺术化的外形在灯光的映衬下，既有工业建筑的美感，又兼具艺术美学的气质，无不呈现出岭南现代建筑的地标形象。

The second phase of Guangzhou Fifth Resource Thermal Power Plant project in Huadu District, Guangzhou City is designed with the concept of “High mountains and Running water”. The whole factory building adopts “de-industrialization” design. The artistic shape is reflected by the lighting.The beauty of the art, combined with the temperament of artistic and industrial aesthetics, all present the landmark image of Lingnan modern architecture.

横琴新区区域供冷站和冷冻水管网工程第三供冷站

District Cooling Station & Cooling Pipe of Hengqin New Area NO.3 Cooling Station

项目位置：广东省珠海市
设计时间：2015 年
竣工时间：2018 年
建筑规模：9,090 m²

Location: Zhuhai, Guangdong
Design: 2015
Completion: 2018
GFA: 9,090 m²

区域供冷供热系统是横琴新区建设的一项市政配套基础设施，该系统向新区的各类建筑提供空调系统的冷源、冬季采暖及生活热水的热源，是横琴岛三联供项目中重要的一环。第三供冷站采用多能源大温差系统，利用吸收机上游电制冷，制取大温差冷冻水，并设有冰蓄冷系统；尖峰供冷能力为 3.04 万 RT，蓄冰量为 6.4 万 RTH；尖峰供热能力为 10.2MW；第三供冷站主要服务区域为口岸服务区及综合服务区，服务半径约 2km，服务总建筑面积约 280 万m²。

District cooling and heating system is a municipal facilities for Hengqin New Area. This system provides cool and heat source for heating, Ventilation and Air Conditioning(HVAC) system and domestic hot water for all kinds of buildings, and is also an important component of Combined Cooling Heating and Power(CCHP) for the new area. Multi-energy and large temperature difference system are adopted in No.3 Cooling Station, chilled water with large temperature difference is prepared by connecting LiBr absorption and vapor compression chiller in series, and ice-storage system is also set in this system. The peak cooling capacity is 30,400 RT, and cool storage capacity is 64,000 RTh. The peak heating capacity is 10.2 MW. Main servant region are the port service area and integrated service area. The service radius is about 2 kilometers and the region is about 2.8 million square meter.

韶关武江乡村振兴传习所

Rural Vitalization Training Center of Wujiang, Shaoguan

项目位置：广东省韶关市
设计时间：2018 年
竣工时间：2021 年
建筑规模：6,000 m^2

Location：Shaoguan, Guangdong
Design：2018
Completion：2021
GFA：6,000 m^2

韶关武江乡村振兴传习所项目在设计构思上充分挖掘场地承载的历史文脉，深度领悟“粮仓”价值意义，阐释粮仓遗产价值，希望通过对粮仓功能解构重组，让粮仓过去的辉煌得以在今日重现，打造一个集历史体验、农产展销、党建宣传于一体的教育示范基地。

基于对原有场地保留历史粮仓建筑的总体布局的尊重，保留了原有的肌理，通过场地地面铺装设计和新建建筑，让老建筑更加凸显出其原本的特色，焕发出其历经时间沉淀后的特殊气质；新建建筑融合了老建筑的形体特征，运用现代的手法和当地材质进行设计，创造出丰富独特的场所氛围。

Rural Vitalization Training Center of Wujiang, Shaoguan fully digs out the historical context of the site, while we understand and explain the value of granary. We aim to reconstruct the function of the granary so that the past brilliance of the granary can be reappeared today, so as to build a set of historical experience, agricultural products sales and the party propaganda, which integrates education demonstration base.

Based on the respect for the overall layout of the original site to retain the historical granary building, the design retains the original texture, which highlights the characteristics of the old building, radiates its special temperament after time precipitation through the site ground pavement design and new-built buildings. The new building, integrates the structure features of old buildings, uses modern techniques and local materials in the design, creating a rich and unique place atmosphere.

侵华日军南京大屠杀遇难同胞纪念馆周边地区城市设计
广州国际金融城起步区城市设计
河北雄安新区起步区城市设计
河北雄安新区启动区城市设计
前海公共空间设计
“河港杯”首届河北国际城市规划设计大师邀请赛——秦皇岛·百年老港转型复兴
中国（海南）自由贸易区海口江东新区概念规划方案
中国（海南）自由贸易区海口江东新区起步区城市设计方案
广州南站地区城市设计优化及控制性详细规划
兵团红星市中心城区城市设计与控制性详细规划
长沙南部片区概念规划
怀柔科学城总体城市设计
惠东考洲洋保护与开发概念性规划
广州科学城提升规划设计
石家庄正定新区中心公园周边地区综合设计
凤凰山景观塔及周边城市设计
广州南沙新区亭角地区城市设计
西安高新区西太路核心区未来科技城特色小镇概念规划
南昌市经开区白水湖片区概念性城市设计

08 城市设计
URBAN DESIGN

Urban Design of the Surrounding Area of the Memorial Hall of the Victims in Nanjing Massacre by Japanese Invaders
Urban Design of the Boot Area of Guangzhou International Financial City
International Consulting for the Urban Design of the Boot Area for the Xiong'an New Area in Hebei Province
The Solicitation of Proposals for the Urban Design of the Boot Area of the Xiong'an New Area in Hebei Province
Qianhai Public Spaces Design
"HPG Cup" the First Hebei International Urban Planning and Design Master Competition-Qinhuangdao Century-old Port Transformation and Revival
Conceptual Planning Scheme of Haikou Jiangdong New District in China (Hainan) Free Trade Area
Urban Design Plans for Haikou Jiangdong New District Starting Area in China (Hainan) Free Trade Area
Urban Design Optimization and Controlled Detailed Planning of Guangzhou South Railway Station Area
Urban Design and Control Detailed Planning of the Corps Red Star City Center
Conceptual Planning of the Southern Region, Changsha
Urban Design of Huairou Science City
Conceptual Planning of Huidong Kaozhouyang Conservation and Development
Promotion Planning of Guangzhou Science City Area
Shijia zhuang Zhengding New District Central Park Area Comprehensive Design
Urban Design of Fenghuang Mountain Landscaped Tower and Surroundings
Urban Design of Tingjiao Area, Nansha New Area, Guangzhou
Conceptual Planning of the Characteristic Town of Future Science and Technology City in the Core Area of Xitai Road, Xi'an High-tech Zone
Conceptual Urban Design of Baishui Lake Area in Economic Development Zone of Nanchang

侵华日军南京大屠杀遇难同胞纪念馆周边地区城市设计

Urban Design of the Surrounding Area of the Memorial Hall of the Victims in Nanjing Massacre by Japanese Invaders

项目位置：江苏省南京市
设计时间：2014 年
竣工时间：2015 年
建筑规模：3,300,000 m^2

Location：Nanjing, Jiangsu
Design：2014
Completion：2015
GFA：3,300,000 m^2

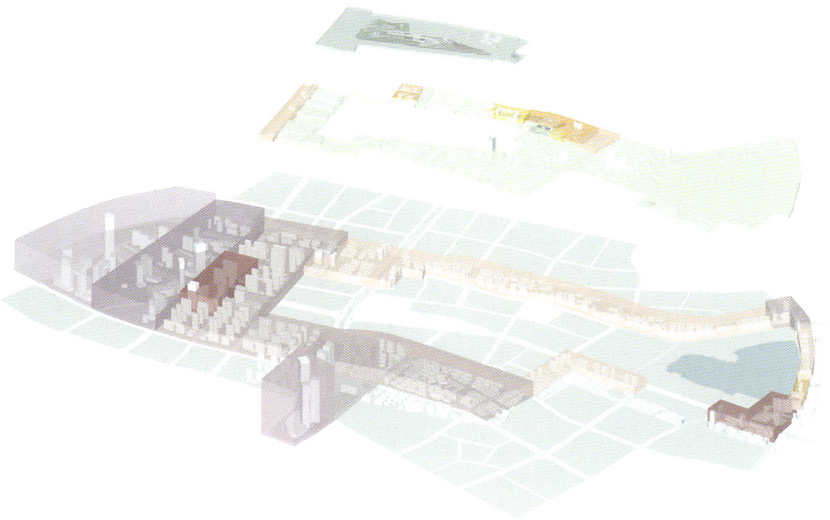

本次城市设计将国家公祭作为核心思想统筹场地内各种功能构成，从现实问题出发，考虑各种可能性，制定最优化目标的同时，提出高适应性阶段目标。通过调整功能、改善存量、优化增量的方法，实现项目所在地的全面提升和优化。在功能细化中，分圈层、分强度控制，严格挑选纪念馆周边地区可承载的功能。

This urban design takes the national public memorial as the core concept to coordinate the various functions in the venue, starts from the realistic problems and considers various possibilities to formulate the optimal goal and proposes the stage goal of high adaptability. Through adjusting functions, improving the stock and optimizing the increment, the overall improvement and optimization of the project sites are achieved. In the function refinement, the possible functions that can be carried around the memorial hall are strictly selected according to their layers and intensity control.

广州国际金融城起步区城市设计
Urban Design of the Boot Area of Guangzhou International Financial City

项目位置：广东省广州市
设计时间：2012—2014 年
建筑规模：4,986,900 m^2

Location：Guangzhou, Guangdong
Design：2012-2014
GFA：4,986,900 m^2

规划按照“低碳经济、智慧城市、幸福生活”的新型城市化发展要求，将广州国际金融城起步区打造成为国内领先的金融集聚区，立足广州，依托珠三角，面向东南亚，打造新型城市化最佳实践区、国内领先的金融集聚区、具岭南特色的中央活力区、国际一流的生态理想城。

According to the new urbanization development requirements of “low-carbon economy, intelligent city and happy life”, the plan aims to build the starting area of Guangzhou International Financial City into the leading financial agglomeration area in China based on Guangzhou, relying on the Pearl River Delta, facing Southeast Asia. It is to create a new type of best urbanization best practice area, leading financial agglomeration area in China. Lingnan characteristic central active area and international first-class ecological city.

河北雄安新区起步区城市设计

International Consulting for the Urban Design of the Boot Area for the Xiong'an New Area in Hebei Province

项目位置：河北省雄安新区

设计时间：2018 年

建筑规模：198 km^2

Location：Xiong'an, Hebei

Design：2018

GFA：198 km^2

华南理工大学建筑设计研究院有限公司

设计方案“淀水方城”有四大规划构思：

生态修复：梳理河道，畅通外水，截污减排，整治淀区，修复湖岸和退耕还淀。考虑季节性差异，以生态手段为调节方式，应对雨季和旱季的不同地表水量，最大化进行生态雨水收集处理。

城淀共融：规划形成“一带、一轴、一湾”的空间结构，“城—镇—村”三级形态系统，以及“高铁枢纽与中轴组团、城市湖岸滨水区、产业特色小镇、文化”四大重点地段。

雄安风格：借鉴中国传统方院与方城格局，形成城市与自然和谐共生的城镇网络。建筑设计通过营城理念的再现、飞檐反宇的现代演绎及园林空间的诗意表达进行风貌控制，形成“中而新”的雄安风格。

未来城市：体现人类对未来美好人居环境的愿景，弱化功能分区，强调多元混合与街区式布局。以人民为中心，形成“城市功能区”“特色小镇区”与“文化生态村落区”的产业发展平台。通过中轴及主要公共空间结合景观水系形成连续的地下活力连廊，实现冬日“地下暖城”的宜居目标，构筑体现中国未来城市的范型。

The four major planning concepts of “Dianshui Square City” as below:

Ecological restoration: Combing river channels, unblocking external water, intercepting pollution and reducing emissions, rectifying the lake area, recovering the lakeside, and returning farmlands to lakes. Considering seasonal differences, ecological means should be taken as the regulation method to deal with different surface water capacity in rainy and dry seasons and maximize ecological rainwater collection treatment.

City and lake coexistence: It is planned to form the spatial structure of “one Belt, one Axis, and one Bay”, the three-level morphological system of “city-town-village”, and four key areas of “high-speed railway hub and central axis group, urban lakeside waterfront area, town with industrial characteristics, and culture”.

Xiong'an style: Drawing on the traditional Chinese square courtyard and square city pattern, a town network with harmonious coexistence between city and nature is formed. Through the representation of the traditional concept of city management, the modern interpretation of upward tiles cornices and the poetic expression of garden space, the architectural design controls the scenes and forms the “traditional and new” Xiong'an style.

Future City: It reflects the vision of human beings for a better living environment in the future, weakens functional divisions, and emphasizes multi-mixed and street-style layouts. Taking the people as the center, it will form an industrial development platform including “urban functional area” “characteristic town area” and “cultural ecological village area”. By combining the central axis and main public space with the landscape water system, a continuous underground dynamic corridor is formed to achieve the livable goal of “underground warm city” in winter and to build a model for China's future cities.

河北雄安新区启动区城市设计

The Solicitation of Proposals for the Urban Design of the Boot Area of the Xiong'an New Area in Hebei Province

项目位置：河北省雄安新区
设计时间：2018 年
建筑规模：38.31 km^2

Location：Xiong'an, Hebei
Design：2018
GFA：38.31 km^2

雄安启动区"淀谷暖城"蕴含四大主题：创造人性宜居的城淀结构，构建活力复合的蓝绿基底，营造冬暖夏凉的创新组团，展现传承文化的雄安风格。城市以小尺度街区和人性化居住空间为结构单元，打造丰富多样的城市生活圈，塑造步行友好，城淀、城绿交融的"宜居淀城"。景观系统强调优化现状水文水系，结合城市雨洪管理与居民日常生活，形成生态复合的水系统和连贯多元的城市景观绿脉，以此构建活力生态的"蓝绿基底"。

Xiong'an Boot Area "Livable Towns" is envisioned around four main themes, to create livable towns, construct an eco-matrix; to cultivate a 26 ℃ City idea and express the contemporary Chinese Xiong'an style. The city is created based on a framework that is composed of human scale streets and livable units. The framework promotes a highly diverse, pedestrian friendly and ecological sound living town idea. The development of the landscape framework emphasizes on the existing hydrological system, integrating with the city-wide water management strategies to generate a comprehensive ecomatrix.

前海公共空间设计

Qianhai Public Spaces Design

项目位置：广东省深圳市
合作单位：北京土人城市规划设计股份有限公司、
东南大学建筑设计研究院有限公司
设计时间：2019 年
建筑规模：112,000 m^2

Location：Shenzhen, Guangdong
Partner：Turenscape, Architects & Engineers Co.,Ltd. of Southeast University
Design：2019
GFA：112,000 m^2

深圳前海合作区是国家“一带一路”的战略支点、粤港澳大湾区创新合作的新核心引擎，将成为面向未来、独一无二的国际级湾区城市新中心。项目面临片区开发强度大、空间压迫、需求多元、慢行安全和炎热多雨的城市设计挑战。设计以“绿腔”为核心概念，整合城市公共空间与城市建筑、环境、交通、人群需求的关系，构建一体化的、功能复合的、立体多元的城市标志性公共空间。此次项目方案由国内领先的城市及建筑设计、景观设计和桥梁设计公司强强联合，三家设计单位的建筑师、景观设计师、桥梁设计师将在前海展开广泛而密切的跨界合作，共同打造前海的城市复合空间名片。

Qianhai Cooperation Zone is the strategic point of the Belt and Road and a new core engine of Guangdong-Hong Kong-Macao Greater Bay Area. It will become a unique new city center of international bay area facing the future. The project faces the challenges of strong development, space oppression, diversified demands, slow-moving safety and hot and rainy weather. The design takes "green cavity" as the core concept, integrating the relationship among urban public space and buildings, environment, transportation and people demands, in order to build an integrated, multi-functional and multi-dimensional symbol of urban public space. This project was jointly led by the leading domestic companies of urban and architectural design, landscape design and bridge design. The architects, landscape designers and bridge designers of the three companies will conduct extensive and close cross-border collaboration in Qianhai, creating a symbol of urban composite space of Qianhai together.

“河港杯”首届河北国际城市规划设计大师邀请赛——秦皇岛·百年老港转型复兴

“HPG Cup” the First Hebei International Urban Planning and Design Master Competition–Qinhuangdao Century-old Port Transformation and Revival

项目位置：河北省秦皇岛
设计时间：2018 年
建筑规模：8.90 km^2

Location: Qinhuangndao, Hebei
Design: 2018
GFA: 8.90 km^2

规划以“汇通海岳 · 连城拥湾”定位西港区，将其作为城市核心资源打造，与东港区互补协同发展，再造秦皇岛城市中央商务区、城市旅游集散地。

空间设计以“一轴、三圈、四象圈层发展”为框架，依托中轴，形成串联山、城、湾、海的城市功能通道，以海洋为原点，形成滨海休闲圈—综合服务圈—城市生活圈三大圈层；打造北城（宜居休闲城）、南湾（魅力国际湾）、东岛（历史文化岛）、西港（邮轮活力港）四象。

考虑北方气候特点，以“冬日暖城 · 北方明珠”为指导思想，通过各式交通无缝衔接公共中心，构建地上地下一体化的全季候步行环境。

重点地区设计以一条 4.5km 长的多彩滨海游憩带串联起西港、南湾、东岛，创造城市型活力水岸；依托开敞水面、滨海岸线构筑港口博物馆、海洋文化馆、媒体中心等标志建筑，创造活力水城共融空间，重新定义海湾格局。

借百年老港复兴契机，秦皇岛市将迎来城市活力的新生，西港湾区将成为一个新时代高质量发展的时空港湾。

West Port District is positioned as the “connecting sea and mountain, linking cities and closing to bay”, which is build as the core resource of the city. It cooperates with East Port District for complementary and coordinated development to re-create Qinhuangdao's urban central business district and urban tourism distribution center.

The spatial design of “one-axis, three-circle, four-image circle development” is taken as the framework. The urban function channel connecting mountains, cities, bays and seas is shaped by relying on the central axis; three major circles of coastal leisure circle, comprehensive service circle and urban life circle are established while taking the ocean as the origin; and four images including north city (livable leisure city), south bay (charming international bay), east island (historical and cultural Island), and west port (cruise vitality port) are built.

Considering the characteristics of the northern climate, under the guiding ideology of “warm winter city·northern pearl”, the public center is seamlessly connected through multiple transportation to construct an all-season walking environment with ground and underground integration.

The key area is designed with a 4.5-kilometer-long colorful coastal entertainment belt connecting west port, south bay and east island to set up an urban dynamic waterfront; relying on the open water surface and the coastline, landmark buildings including the port museum, marine culture museum, media center are set up to create a vibrant water city and redefine the pattern of the bay.

Taking the opportunity of the revival of the century-old port, Qinhuangdao will usher in a new birth of urban vitality, and West Port District will become a time-space harbor with high-quality development in the new era.

中国（海南）自由贸易区海口江东新区概念规划方案

Conceptual Planning Scheme of Haikou Jiangdong New District in China (Hainan) Free Trade Area

项目位置：海南省海口市
设计时间：2018 年
建筑规模：418 km^2

Location: Haikou, Hainan
Design: 2018
GFA: 418 km^2

本项目以海促城、以水营城，打造生态、零碳、独具特色的亚热带滨海城市典范；以港兴城、以产融城，建设开放包容、共享、深化改革开放的国际交往城市先锋。以生态优先为原则，规划提出蓝绿交织、城水相依的指状空间，生态与城市咬合的结构。指状生态结构以大海为出发点，形成多条指状蓝绿脉络，联通海、田、林、湖、江，成为网络化的生态框架。以空港为掌，指状岛链结构，择高地及优化原有建设区，形成五条城市发展轴，承接国际、国家、以及海口市重要功能，激活空港流量，发展未来产业。

This project, promoting the city with the sea and managing the city with waters, aims to create an ecological, zero-carbon and unique subtropical coastal city model and to build a pioneering international exchange city that is open, comprehensive, shared and dedicated to deepening reform and opening up. Based on the principle of ecological priority, the plan proposes a blue-green intertwined and urban-water-dependent finger-shaped space, an occlusive structure of ecology and city. The finger-shaped ecological structure takes the sea as the starting point and forms a plurality of finger-shaped blue-green veins, connecting the seas, the fields, the forests, the lakes and the rivers into a networked ecological framework. With the airport as the palm and a finger-chain structure island, the high ground is selected and the original construction area is optimized to form five urban development axes, taking on important functions for the world, the country and Haikou city, activating airport traffic and developing future industries.

中国（海南）自由贸易区海口江东新区起步区城市设计方案

Urban Design Plans for Haikou Jiangdong New District Starting Area in China (Hainan) Free Trade Area

项目位置：海南省海口市
设计时间：2018—2019 年
建筑规模：1.79 km^2

Location：Haikou, Hainan
Design：2018-2019
GFA：1.79 km^2

江东新区承载了中国（海南）自贸区（岛）的集中展示任务，是海南新时代全面深化改革开放的新标杆，展示中国风范、中国气派、中国形象。起步区是江东新区的先行开发区之一，本项目面临的挑战是如何在有限场地内建设一个韧性安全、环境优美、功能完善的城市。我们以生态为底，安全韧性为原则，规划坚持共生、共融、共享的规划理念，从“凉城、融城、活城”三条路径出发，为起步区构想一个全天候适应、全要素聚合、全时段交往、全方位创新的总部新城。

Jiangdong New Area, carrying the concentrated display task of China (Hainan) Free Trade Zone (Island), is the new benchmark for Hainan's comprehensively deepening reform and opening up in the new era, showing China's demeanor, manner and image. The starting area is one of the pioneering development zones of Jiangdong New District. The project faces how to build a city with resilient safety, beautiful environment and perfect functions in a limited space. Based on the principle of ecology and resilient safety, adhering to the conception of symbiosis, integration and sharing considering the three paths of "cooling the city, integrating the city and vitalizing the city", an all-weather adaptation, all-element aggregation, full-time contact and all-round innovative new headquarters city is envisioned for the starting area.

广州南站地区城市设计优化及控制性详细规划

Urban Design Optimization and Controlled Detailed Planning of Guangzhou South Railway Station Area

项目位置：广东省广州市
设计时间：2013—2016 年
批复时间：2016 年
建筑规模：36.16 km²

Location: Guangzhou, Guangdong
Design: 2013-2016
Approval time: 2016
GFA: 36.16 km²

规划构思：规划整合南站地区交通门户、山水空间、地域特色、时代精神等资源，重新审视上一轮规划设计方案，本次规划提出六大策略：①文化策略——岭南风貌·山城水韵；②功能策略——公服引领，价值提升；③高效交通——空铁联运，高效集散；④地下空间——核心商业·复合利用；⑤空间形态——岭南神韵，疏密有致；⑥旅游策划——生态文化·综合服务。最终形成更具有整体性和可持续发展的城市空间发展框架。

空间结构形成“一心、两轴、三环、六片区，圈层式辐射发展”的整体格局：

一心：以广州南站交通枢纽为核心，形成高度聚集和高效的商务商贸服务核心。

两轴：以广州南站为核心，依托汉溪大道和兴业大道打造城市向外辐射的主轴线。

三环：形成山水生态休闲环、岭南文化活力环、公共服务环。

六片区：枢纽核心区、预留发展区、东部发展区、石壁发展区、屏山发展区、沙湾发展区六大功能区。

Planning Conception: After integrating the traffic gateways, landscape space, regional characteristics, spirit of the times and other resources in Guangzhou South Railway Station area and re-examining the previous round of planning and design schemes, this plan puts forward six strategies as below: ① cultural strategy-Lingnan style landscape of mountains and rivers; ② functional strategy-led by public services to enhance value; ③ efficient transportation-air-rail transport and efficient distribution; ④ underground space-core commerce and compound utilization; ⑤ spatial form-Lingnan charm, dense and compact; ⑥ tourism planning-ecological culture and integrated services. The six strategies ultimately forming a more holistic and sustainable urban space development framework.

The spatial structure is the overall pattern of "one center, two axes, three rings, six zones, and circle and layer radiation development".

One center: taking Guangzhou South Railway Station as the core, a highly concentrated and efficient business and trade service core is formed.

Two axes: taking Guangzhou South Railway Station as the core and relying on Hanxi Avenue and Xingye Avenue, the main axis of urban outward radiation is created.

Three rings: landscape ecological Leisure ring, Lingnan cultural vitality ring and public service ring are formed.

Six zones: the six key functional zones include hub core zone, reserved development zone, eastern development zone, Shibi development zone, Pingshan development zone and Shawan development zone.

兵团红星市中心城区城市设计与控制性详细规划

Urban Design and Control Detailed Planning of the Corps Red Star City Center

项目位置：新疆维吾尔自治区红星市
设计时间：2015—2016 年
建筑规模：32.57 km^2

Location：Hongxing, Xinjiang
Design：2015-2016
GFA：32.57 km^2

城市设计延续地域性景观，将中心城区南部的沙漠作为重要的地域性元素进行保留，沙漠优良细滑的沙质也成为未来发展都市沙漠旅游的重要基础。以天山和沙漠为核心组织地域性景观体系，建立北望天山、南临沙漠的城市中央生态轴线，成为展现绿洲风貌和都市活力的生态核心。城市各组团的建设中，突出"看得见天山"作为居民心理方向的指引作用，通过合理布局广场、绿地、道路等开敞空间，使各个城市组团都能够通过视线通廊北望天山，形成极富地域性形象与特色的城市空间。

The urban design continues with the regional landscape, preserving the desert in the southern part of the central city as an important regional element and the fine and smooth sands of the desert as an important foundation for the development of urban desert tourism in the future. Establishing the regional landscape system in Tianshan Mountains and the desert, building the city's central ecological axis that faces Tianshan Mountains in the north and closes to the desert in the south, which becomes the ecological core that displays the oasis and urban vitality. In the construction of various groups of the city, the "visible Tianshan" is highlighted to play the guiding role of the residents'psychological direction. Through the rational layout of open spaces such as squares, green lands and roads, each city group can see the Tianshan Mountains in the north through view corridors, forming a city rich in regional images and characteristics.

长沙南部片区概念规划

Conceptual Planning of the Southern Region, Changsha

项目位置：湖南省长沙市
设计时间：2018 年
建筑规模：293.89 km^2

Location：Changsha, Hunan
Design：2018
GFA：293.89 km^2

规划构思：“智谷绿心”（人才交流与休闲中心，创新生态发展高地）；“负阴而抱阳，坐绿而拥城”（实现生机永续的绿色城市）。

在紧凑高效的前提下，尊重现有的城市建设与乡村交错的空间格局，创造由若干城、镇、村散落林间水畔的独特布局。以轨道交通站点为枢纽，沿南北主轴和东西主轴交汇布局城，在生态过渡区布局镇；避开森林、林地，改造村落现状，在生态核心区域布局村，形成城镇村三级复合城市群落。主动有效地控制城区轮廓，增加绿色边界，使土地价值最大化。合理设定开发强度，形成主题鲜明的特色区域。

空间结构沿“一带两轴一廊”有机展开，中部融城发展轴依托轨道站点集中布置区域城市职能、公服设施、科创与文创产业及居住功能；东西创新发展轴，将交通枢纽与滨水生活进行连接。将湘江风光带—智谷绿心—东部生态绿心廊道打通，建造城市绿肺，结合发展健康养生、生态休闲、文创旅游等生态绿色产业，振兴美丽乡村。

Planning Idea: “Wisdom Valley and Green Center” (talent exchange and leisure center, innovative ecological development highland); “Combination of Yin and Yang, coexistence of green space and city” (a green city with sustainable vitality).

Under the premise of compactness and high efficiency, on the basis of respecting the existing urban construction and the spatial pattern of rural interlacing, a unique layout with several cities, towns and villages scattered among the trees and by the water is created. The rail transit station is taken as the hub to layout the cities along the north-south main axis and the east-west main axis, and layout the towns in the ecological transition zone. By avoiding forests and wood lands, existing villages are renovated in ecological core area, forming a three-level composite city community including cities, towns and villages. The planning controls the urban contour actively and effectively, increases the green boundary to maximize the land value, and reasonably sets the development intensity to form distinctive characteristic areas with distinctive themes.

The spatial structure is organically developed along the “one belt, two axis, and one corridor” . The center of the Rongcheng relies on the stations to centrally arrange regional urban functions, public service facilities, scientific innovation and cultural creative industries and residential functions; and the east-west innovation development axis connects the transportation hub with waterfront life. For the Xiangjiang Scenic Belt-wisdom valley and green center-eastern ecological green center corridor is opened up and has become the urban green lung. The ecological green industries such as healthy care, ecological leisure, an the cultural creative tourism are also developed to revitalize beautiful countryside.

怀柔科学城总体城市设计

Urban Design of Huairou Science City

项目位置：北京市
设计时间：2018 年
建筑规模：100.96 km^2

Location：Beijing
Design：2018
GFA：100.96 km^2

目标定位于将怀柔打造成一个自然和城市平衡发展的科学新城，一个激发全球科技创新、经济高速增长的引领之地、一个体现文化自信和文化传承的人才聚集地。规划以生态安全为本底，构筑以科学城为核心，区域联动发展的空间格局，通过建立产业、交通及内部空间的联系，整合并带动周边地区协同发展，实现怀柔科学城区域空间资源环境的合理配置、统筹与发展。

The design has been positioned to make Huairou a new science city with a natural and urban balanced development framework, a leading place stimulating global scientific and technological innovation and rapid economic growth, a talent-gathering place that reflects cultural confidence and cultural heritage. Based on ecological security, the plan intends to construct a spatial pattern with science city as the core and develop with regional linkage. By building connections of industry, transportation and inner space, the design integrates and drives the coordinated development of surrounding areas, realizing the rational allocation, co-ordination, and development of areas, spaces, resources and environment in Huairou Science City.

惠东考洲洋保护与开发概念性规划

Conceptual Planning of Huidong Kaozhouyang Conservation and Development

项目位置：广东省惠州市
设计时间：2013 年
建筑规模：389 km^2

Location: Huizhou, Guangdong
Design: 2013
GFA: 389 km^2

基于对考洲洋发展趋势的认知和核心价值的选择，规划将考洲洋打造成惠州国际健康产业基地，珠三角东岸重要的养生胜地，全球生命科学产业集群的标杆地以及人类候鸟式居游的目的地。本次整体开发思路为以保护生态为目标，建立生态环洋链珠体系，形成生态保护框架；创造“以健康为核心”的产业体系，打造品牌价值文化，形成带动地区可持续发展的经济框架；深入挖掘城市环境内涵，成为人居环境典范。

Based on the cognition of development trend and the choice of core values for Kaozhouyang, Kaozhouyang will be built as the Huizhou international health industry base, an important health resort on the east coast of the Pearl River Delta, a standard for the global life science industry cluster and a destination of residence and tour like migratory birds. The overall development idea is: with the goal of protecting ecology, establishing an ecological circle chain system to form an ecological protection framework; creating an industrial system with "health as the core" to establish a brand value culture and form an economic framework for driving regional sustainable development; excavating deeply into the connotation of urban environment and becoming a model for human settlement environment.

广州科学城提升规划设计
Promotion Planning of Guangzhou Science City Area
项目位置：广东省广州市
设计时间：2019 年
建筑规模：144.6 km²
Location：Guangzhou, Guangdong
Design：2019
GFA：144.6 km²

在粤港澳大湾区的发展战略方针下，广州科学城抓住机遇，协同广州中新知识城、广州国际创新城、琶洲互联网集散区齐头发展，打造粤港澳大湾区的开放型的全球创新网络中心，融入综合性的国家科学中心，成为引领未来发展的中国智能制造中心以及制度创新先行区。本次规划尊重场地山水特征，构建连山通水的三大城市功能服务带。沿道路及绿地形成串联组团中心的城市活力共享链，集中布置商业与科创服务设施。最终形成"一链三带"的城市空间格局，建设连通、集聚、高效、绿色的公园城市。目的是将广州科学城建设成为粤港澳大湾区国际创新中心的"湾顶明珠"。

Under the development strategy of Guangdong-Hong Kong-Macao Greater Bay Area, Guangzhou Science City seizes the opportunity, cooperates with Guangzhou Zhongxin Knowledge City, Guangzhou International Innovation City and Pazhou Internet Distribution Center to develop an open global innovation network center, integrate with comprehensive national science center, to become China intelligent manufacturing center leading future development and an institutional innovation pioneer district at Guangdong-Hong Kong-Macao Greater Bay Area. This plan constructs three major urban functional service belts connecting mountain and water on the basis of respecting the characteristics of landscapes. Along the roads and green lands, a series of urban vitality sharing chains are formed to centrally arrange commercial and scientific innovation service facilities. Finally, the urban spatial pattern of "one chain and three belts" is shaped to build an connected, clustered, efficient and green park city. Eventually, Guangzhou Science City will be built into "the pearl at the top of bay"of the International Innovation Center of Guangdong-Hong Kong-Macao Greater Bay Area.

石家庄正定新区中心公园周边地区综合设计

Shijianzhuang Zhengding New District Central Park Area Comprehensive Design

项目位置：河北省石家庄
设计时间：2011 年
竣工时间：2015 年
建筑规模：2.8 km^2

Location: Shijiazhuang, Hebei
Design: 2011
Completion: 2015
GFA: 2.8 km^2

规划定位为“正定之心——城市原点、城市客厅与永续传承的精神地标”，形成古今对话、十字发展与疏密有致的城市格局。营造文化与生态共生的核心公共空间，塑造起步区的门户形象，强化起步区的带动效应，激发滨河的城市活力。通过“一心一网，三轴两带，十大片区”的空间结构，打造滨水魅力之城、创新动力之城和宜居活力之城。

The plan position is positioned as “the heart of Zhengding District—spiritual landmark of the urban origin, the urban living room and the sustainable inheritance”. It forms an urban pattern with ancient and modern styles, cross development and proper density. The design establishes a core public space where culture and ecology coexist, shapes the portal image of the starting area, strengthens the leading effect of the starting area, and inspires the riverside urban vitality. Through the spatial structure of “one heart and one network, three axes and two belts, and ten large sectors”, the city is characterized with riverside charm, innovation, and livability & vitality.

凤凰山景观塔及周边城市设计

Urban Design of Fenghuang Mountain Landscaped Tower and Surroundings

项目位置：广东省珠海市
设计时间：2014 年
建筑规模：161,459 m²

Location: Zhuhai, Guangdong
Design: 2014
GFA: 161,459 m²

这是为珠海市在凤凰山区域规划设计的城市公园以及文化地标。我们不仅希望通过规划设计，为市民营造良好的户外休闲公园，更希望这是一个展现珠海城市形象、凤凰山景观特色的文化地标。

场地的景观路径等元素以梧桐树的枝茎为原型，凤凰塔及周边重要节点的形态以“百鸟朝凤”为寓意，赋予场地灵力，营造城市休闲公园，塑造城市文化地标。

This city park and cultural landmark was designed for Zhuhai's Fenghuang Mountain. We imagined not only to create an outdoor leisure park for citizens, but also make it a cultural landmark to show the image of Zhuhai and Fenghuang Mountain to the world.

The landscape is based on the branches and stems of sycamore tree, while the Fenghuang Tower is designed in the form of “birds paying homage to the phoenix”. This grants the site the power of wisdom, and creates an urban leisure park that will become a new landmark for the city.

广州南沙新区亭角地区城市设计

Urban Design of Tingjiao Area, Nansha New Area, Guangzhou

项目位置：广东省广州市
设计时间：2017 年
核心区面积：6.8 km²
建筑规模：13.2 km²

Location：Guangzhou, Guangdong
Design：2017
Core Area：6.8 km²
GFA：13.2 km²

南沙是国家级新区、国家自贸试验区、国家自主创新示范区，既承担制度创新和加快发展的重任，同时也承担着高水平对外开放的门户枢纽、广州城市副中心的重任。亭角地区是进入自贸区南沙新区和城市中心区的门户，对体现广州城市副中心形象具有十分重要的意义。设计通过改造原有旧厂房，置入高品质文化馆，通过旧厂转型、产业再造达到文化引爆的效果，将南沙从工业海港改造成为创意人才港。

Nansha is a national new district, a national free trade pilot zone, and a national independent innovation demonstration zone. It not only undertakes the important task of institutional innovation and accelerated development, but also undertakes the task of running a high-level open door to the hub and the Guangzhou city sub-center. Tingjiao is the gateway of the Nansha New District and the urban center of the free trade zone. It is of great significance to reflect the image of the Guangzhou city sub centre. The design reforms the old factory building into a high-quality cultural center, transforms the industrial port into a creative talent port through the transformation of the old factory and industrial re-engineering to achieve the effect of cultural detonation.

图书在版编目（CIP）数据

华南理工大学建筑设计研究院有限公司作品选：2009—2019：汉英对照 / 丘建发，裴泽骏编著．—广州：华南理工大学出版社，2021.12
ISBN 978-7-5623-6522-8

Ⅰ．①华… Ⅱ．①丘… ②裴… Ⅲ．①建筑设计－作品集－中国－现代 Ⅳ．① TU206

中国版本图书馆 CIP 数据核字（2020）第 218150 号

华南理工大学建筑设计研究院有限公司作品选（2009—2019）
丘建发 裴泽骏 编著

出 版 人：卢家明
出版发行：华南理工大学出版社
（广州五山华南理工大学 17 号楼，邮编 510640）
http://hg.cb.scut.edu.cn E-mail: scutc13@scut.edu.cn
营销部电话：020-87113487 87111048（传真）
策划编辑：赖淑华
责任编辑：骆 婷
责任校对：梁晓艾
印 刷 者：广州一龙印刷有限公司
开 本：787mm × 1092mm 1/12 印张：26.5 总字数：1326 千
版 次：2021 年 12 月第 1 版 2021 年 12 月第 1 次印刷
定 价：298.00 元